W9-BBH-369

A SHEARWATER BOOK

NO MAN's GARDEN

NO MAN's GARDEN

THOREAU
AND A NEW VISION
FOR CIVILIZATION
AND NATURE

Daniel B. Botkin

ISLAND PRESS / Shearwater Books
Washington, D.C. · Covelo, California

A Shearwater Book published by Island Press.

Library of Congress Cataloging-in-Publication Data

Botkin, Daniel B.
No man's garden : Thoreau and a new vision for civilization and nature / Daniel B. Botkin.
p. cm.
Includes bibliographical references (p.) and index.
ISBN 1-55963-465-0 (cloth : alk. paper)
1. Thoreau, Henry David, 1817–1862. 2. Naturalists—United States—Biography. 3. Authors, American—19th century—Biography. 4. Civilization. 5. Nature. I. Title.
QH31.T485 B68 2001
304.2—dc21
00-010445

Printed on recycled, acid-free paper

Printed in Canada

10 9 8 7 6 5 4 3 2 1

To my sister,
Dorothy B. Rosenthal,
who has been a lifelong
source of inspiration, support,
ideas, and books to read, and is
my harshest and best critic

Contents

"Nature was here something savage and awful, though beautiful. . . . This was that Earth of which we have heard, made out of Chaos and Old night. Here was no man's garden, but the unhandseled globe. It was not lawn, nor pasture, nor mead, nor woodland, nor lea, nor arable, nor waste land. It was the fresh and natural surface of the planet Earth."

Henry David Thoreau,
describing the summit of Mount Katahdin, Maine,
late summer 1846

Introduction

This book is about some of the great, ancient questions that people have always asked, questions that remain central, deep, and pertinent in our time. The first question is, What is and what can be a person's connection with nature, both physical and spiritual? The second is, What is and ought to be the connection between civilization and environment: can both survive together, or must one give way to the other? The third is, How can one make contact with and know nature? In this book, I describe an approach to the knowledge of nature that I believe could be of help with these questions. It is a method apparent in Henry David Thoreau's life and work.

Our idea of nature is forged from ancient myths, persistent cultural mores, contemporary political ideologies, and our individual wishes, dreams, and desires, but it is recast by our contemporary nature-knowledge, which in turn is affected by our technology. As a result, in times of societal or technological transition, the fundamental questions about religion and nature and about civilization and environment must reappear and be resolved anew.

The environmental movement of the early twenty-first century is the current formulation of some aspects of these ancient issues and is, to some extent, the outer crust of steely problems. The term *environmentalism* means different things to different people, but the environmental movement is generally perceived as a political and social movement—sometimes as just another one of those special interests lobbied for on Capitol Hill—and no more. Yet the issues referred to as "environmental" are much deeper. They strike at the heart of our existence and the heart of the human condition, affecting every part of our lives.

The relationship between an individual human being and nature is so important, fundamental, and deep within each of us that we are generally unaware of its significance; it is intrinsic to each of us. We take this rela-

tionship for granted during periods of relative cultural stability, but our awareness of it tends to bubble to the surface in times of technological or other major change. This happened between the seventeenth and mid-nineteenth centuries, during the first scientific-industrial revolution, the revolution of physics and steel. It is happening again today, at the beginning of the third millennium of the common era, as we undergo a second scientific-industrial revolution, the revolution of biology and computers. This revolution began in the second half of the nineteenth century with Charles Darwin's and Alfred Russel Wallace's theory about the origin of species, the theory of biological evolution. It is reaching a new stage with new knowledge about ecology—the relationship here and now between living things and their environment.

These transitions take place in part because the way we think about nature is greatly influenced by the limits of our ability to observe nature and therefore is affected by our technologies. New science leads to new technologies, which lead to new observations. The new observations lead to a new understanding of how nature works. This understanding leads to another set of new ideas, which then lead to another set of new technologies, and the process repeats itself. Each change in ideas about nature alters beliefs about relationships between people and nature and between civilization and nature. The rise of science therefore has profoundly affected human perceptions of nature and of the fundamental questions about humanity's connection with nature.

When such transitions occur, they grind against the wheels of culture, society, and politics. The grinding down of old views proceeds with the same inevitability with which an unoiled bearing wears through. Most people and social institutions resist such a change because it is painful to alter one of the oldest and deepest concerns of human beings, the relationship between people and nature.

This book comes about at a time of renewed polarization of opinion about the environment. At one extreme, proponents of the recent philosophical movement called "deep ecology" argue that our primary moral imperative is to save the biosphere—Earth's planetary life-support system—and that all else, including the size of the human population and the essence of civilization, must be sacrificed for this goal. At the other extreme, a conspiracy theory is moving through our society: that environmentalism, under the direction of the United Nations, is being used to destroy private property rights, capitalism, democracy, national sovereignty, even the United States itself. There is a set of books that deal with the politics of anti-environmentalism, such as Ron Arnold's *Ecology Wars,* referred to as the "bible of the wise use movement," and Andrew Rowell's *Green Backlash.* Some argue that around the world, environmentalism is being done in by public relations firms and heavy-handed tactics including beatings and murder. Each extreme sees the "other side" as evil and dangerous to individuals and the world.

These extremes do not represent the majority of people in the United States. Sociologist Thomas Dietz said that despite the presence of these extreme viewpoints, "members of the general public concerned about the fate of the biosphere are also concerned about the fate of other people."

I call for an approach that differs from the extremes, one consistent with the traditions of Western civilization. This approach assumes that ideas arise from observations—of nature, of civilization, and of the self—including observations of what is important to a person in terms of the human spirit, religious feelings, and creativity. Ideas therefore can change over time with changes in nature, civilization, and technology. Information, knowledge, and understanding form a foundation from which ideas and then beliefs arise. This is a vertical view of our existence, with a foundation in observation.

Those who see the battle as rooted only in politics and society might consider my approach naive and call for more forceful action. People are dying from civilization's effects on the environment, and something must be done about it. There is no time to think, one might believe. But I counter that wars have been waged in the name of beliefs and ideologies for many centuries, and I would like to see a gentler, more humane and civilized path.

This book focuses on Henry David Thoreau. Why? Thoreau is an icon of environmentalism, commonly perceived as the father of the idea that civilization has led us down the wrong path, that we must return to nature and abandon much of civilization to save ourselves and nature. But this is not how Thoreau saw civilization and nature. Rather, Thoreau's life is a metaphor for the search for a path to nature-knowledge and a resolution of the questions inherent in humanity's relationship with the rest of the natural world. He focused on this relationship, keeping detailed journals that expressed his thoughts as they developed through his life. In doing so, Thoreau developed an approach to knowledge that is useful for us today. He was willing, perhaps driven, to consider the spiritual as well as the material element in the relationship between human beings and nature. Because he valued civilization as much as he valued nature, his search can serve as a starting point for us in our efforts to achieve goals that benefit both. Thoreau's approach to knowledge is not unique to him, but his life, thoughts, and feelings beautifully illustrate that approach. I do not suggest that Thoreau had all the answers, but he certainly asked the right questions. His search dealt with questions as old as our species, expressed since the beginnings of art and writing, since the earliest philosophy.

Although this book focuses on Henry David Thoreau, it focuses on him as a metaphor, an example. It is not intended to be a scholarly treatise about Thoreau. Many others who are better trained in history than I am, with greater experience in the study of Thoreau's work, are better suited to that task. Rather, my concern is with humanity and nature in the same ways that these were concerns to Thoreau.

An examination of how one can make contact with and know nature—an epistemology of nature—is consistent with the tradition of rationality in Western civilization. This extends to the combination of the rational and the spiritual as occurred when the Greeks began to interact with the intellectualizing and spiritual cultures of the Middle East. *One can view this book as an exploration of how this Western tradition might play out.* It may be that Thoreau's approach, with the metaphor it provides for us today, is the logical extension of Western civilization. If so, and if one believes in the traditions of Western civilization and takes into account modern scientific knowledge about nature, then one will arrive at the same solutions as did Thoreau.

An alternative is to reject completely the traditions of Western civilization, as deep ecologists suggest we should do. In this context, the investigation I pursue in this book could be phrased as a question: Are *Western* civilization and nature compatible? Can we have a *highly technological* civilization that allows for the continued, sustained, natural functions of biological nature essential to our survival? The proposition of deep ecology—that we must abandon Western civilization—is a fascinating and important new idea that deserves examination. Personally, I hope we do not have to take that route because I like Western civilization. The discussion in this book of deep ecology is *not* an attempt to find a straw man, an opponent with views so extreme that it can be easily knocked down. Rather, it is an attempt to face the question of whether modern society can achieve a connection between the human spirit and nature, and between civilization and nature, that builds on the ideas and tools of Western civilization. This was the path along which Thoreau walked, and he serves to epitomize the issues.

The idea of rejecting Western civilization opens up the possibility of adopting some other civilization and its approach to nature. I do not pursue the connection between non-Western civilizations and nature in this book, for the following reasons: Western civilization is enough of a topic for one book; it is what I know, and I do not pretend to be expert enough on other civilizations and cultures to make a similar analysis of them; and Western civilization is the technological force so often blamed for the condition of the environment today. An analogous discussion of contact between human beings and nature in non-Western civilizations would be a welcome contribution and one I would like to read; however, I am not the person to write it.

Although technology is powerful, civilization, the source of technologies, is fragile, more fragile than most of nature. Bacteriologists believe that after a nuclear war, even if human civilization is destroyed, bacteria will continue to make sugars from light, water, and carbon dioxide; to fix and release nitrogen; to grow, reproduce, and evolve. Civilization is much less likely to survive such a catastrophe. It therefore deserves our concern as much as do endangered species and biological diversity.

Perhaps my deep concern with civilization and nature results from my life experiences. I grew up in a house with some twelve thousand books, my father's private library. My father, Benjamin A. Botkin, was one of the country's authorities on American folklore. His interests were eclectic, and as a child I browsed through books that ranged among all of humanity's ideas, from *English Wayfaring Life in the Middle Ages* to a history of the comic strip, from the multivolumed *Century Dictionary,* which had four pages of tiny type about the word *on,* to the *Best of Science Fiction of the 1950s;* from the poetry of E. E. Cummings in lowercase type to the works of William Shakespeare. There was never a thought that these were without value.

To this house came many visitors from many cultures. With them, we listened to strange kinds of music from around the world. One of our frequent folksinging visitors was Pappy, an old man from Trinidad. Because both of my grandfathers had died, Pappy soon became my surrogate grandfather.

My uncle Harry was an artist. His paintings hung in our living room, where he and I discussed the theory of art, the reasons why he painted his pictures, and the history of twentieth-century painting. In short, civilization and its contributions to human creativity were unquestionably of great value to me. They were never in conflict with the outdoors, where I would go with my high school buddies for overnight hikes through pathless woodlands, making camp and cooking potatoes in fire pits.

Before I entered the profession of ecology, I had already come to love the forests of New England. Working with my father-in-law, Heman Chase, a New Hampshire country surveyor, I surveyed forests, bogs, and uplands in southern New Hampshire and Vermont—areas much like the Maine woods of Thoreau's experience. The more time I spent in the woods, the more I realized that here was the place for me to pursue a career—to study nature, to try to understand what I had come to value so much. Why did this tree grow here? What had formed the landscape we were surveying? Why was there a bog here, a lake there, in the complex New England countryside? I had sought a profession that would combine science and the outdoors, and here was that profession, revealed in the forests of New England.

But the days Heman and I spent surveying involved more than the study of nature. When we stopped for lunch, as we drove to and from work, or in between calling out numbers and compass directions, my father-in-law and I talked about the state of human society and what might be done to improve the lot of human beings and the condition of our civilization. For Heman, the ideal life was one in which he could be self-sufficient, be part of the outdoors, enjoy people, and study the best of civilization—spend the day in the woods and the evening at a classical music concert—but most of all, help other people in their lives and try to leave the world a little better place than he found it. For thirty years, he heated his house with wood that he cut with his own hands on sixty acres he had set aside for production of his firewood, using ten cords per winter. He split the cut logs with a tractor-run rig that he had invented and patented. He built his house with his own hands. A country philosopher, Heman had much to say about the human condition. He wrote his own book on economic philosophy and published it at his own expense. He liked the outdoors, but he also loved people. He was well known in our corner of New England as a storyteller and conversationalist, and people dropped by just to listen to Heman and share their thoughts with him. He published a book of stories about local characters and community events. He valued a life lived close to the land, but he was deeply involved with people and civilization. His was very much a Thoreauvian life.

My time spent in the forests studying nature did not seem in any way opposed to an appreciation of human culture or civilization. I like both nature and civilization and would like to see both continue, not one sacrificed for the other. To me, the debates in recent decades about nature versus civilization and wilderness versus technology seem misdirected, for many reasons. The diversity of life on Earth is one of the wonders of the universe, yet so are the products of the human mind: poetry, song, painting, ideas, the human spirit.

When Aldo Leopold set forth a land ethic in his famous book *A Sand County Almanac,* he wrote, "We can be ethical only in relation to something we can see, feel, understand, love, or otherwise have faith in." He wrote that "the first ethics dealt with the relation between individuals," whereas the second dealt with individuals and society at large. He proposed a third stage of ethics that would deal with the relationship between people and the land.

His proposal has become a powerful idea, but with the unforeseen consequence that some have taken it to be a land *versus* people ethic. This is not my reading of Leopold's intentions. Indeed, he wrote that "conservation is a state of harmony between men and land." What I call for in this book is a land *and* people ethic. It has two components. One is the civilization–ecology ethic; the other is the individual–nature ethic. Just as with Leopold's land ethic, neither of these is possible without love, understanding, and faith—what I refer to as spiritual or religious qualities.

This land *and* people ethic begins with joy, love, and reverence for land and life, people and civilization: a love of nature's beauty and diversity as well as a love of human creativity, curiosity, and innovation, as expressed in music, painting, sculpture, literature, and architecture, including landscape architecture. In short, the land *and* people ethic includes a love of the best of humanity's qualities and achievements. My reading of the works of Leopold and Thoreau is that both men saw the world this way. Some might claim that their approach was possible simply because civilization's worst effects on nature were not yet evident in their times. But to believe this is to fail to recognize how ancient the issues are and how central they have been to civilization; it is to fail to recognize the profound effects that much earlier technologies have had on nature.

I have long been an advocate of an environment that is healthy in the way Leopold defined the term: capable of self-regeneration. I have spent years trying to help solve environmental problems. There is no denying that human persistence on the Earth has had, and continues to have, effects on the environment, some negative. Many thoughtful people have written books about these environmental issues, from George Perkins Marsh's great nineteenth-century work *Man and Nature* to Paul Sears's eloquent *Deserts on the March,* written during the dust bowl era; from the often-cited *Silent Spring* of Rachel Carson, *Sand County Almanac* of Aldo Leopold, and novels and essays of Edward Abbey and Peter Matthiessen to the many books on specific environmental issues. I acknowledge the value and significance of these works and the role they have played in modifying the way our society approaches environmental issues. Without denying the influence of humanity on nature, I am alarmed at the nature-versus-people and people-versus-nature trends that I see persisting, perhaps growing worse, among my colleagues and in the developed countries of the world.

The stresses between extreme points of view are pulling at the heart of our society and seem likely to intensify unless we choose a different path to understanding humanity's connection with nature. I believe that a different kind of book is needed at this time, one that will help modify our approach to an integrated system of civilization and nature and of people within nature. Although some who read this book may take it as simply a criticism of environmentalism and environmental activities, that is not my belief or my purpose. Rather, my purpose is to help adjust our approach to living within nature and to integrating civilization and nature, in the hope that both can prosper and persist. This is a constructive goal, but it requires a careful examination of the way we have approached these topics in recent times.

Industrial Western civilization has been the first to view wilderness in a positive way rather than as something to be feared. It is also the first in which many believe that the construction of a city in a desert is not a triumph of civilization but something to be laughed at, denigrated as a waste of water. In the history of civilization, the latter seems a sad and strange turn of events. In losing our sense of our place in nature, we lose our sense of our civilization as well.

My goal, then, is to find a way to help our civilization recover an appreciation of both humanity and nature. *The issue for a rational, well-meaning, and prudent person is how to increase the probability of the combined persistence of both nature and civilization and to maintain for human beings both physical and spiritual well-being.*

In this book, I propose several pathways to resolving the perceived conflicts between the proponents of civilization and the proponents of nature. Swamps, some of Thoreau's favorite places, offer a helpful metaphor. In a swamp, there is a thin line—just a few inches' difference in elevation—between a hummock so dry that a pine tree will grow on it and the water level itself, where water lilies float among sphagnum moss, pitcher plants wait to trap wayfaring insects, and cedars dip their roots into the water. Life permeates these swamps. The water that flows out is tea colored, dyed with tannins from the hundreds of thousands of leaves that fall into the swamp.

Different forms of life have adapted to the subtle differences in elevation that create hummock and open water. In the open water are flowering plants that grow beneath the surface or, like the water lily, root at the bottom but float leaves on the surface. Along the edge between hummock and open water grow emergent flowers, sedges, and shrubs, their roots able to survive in the oxygen-scarce submerged soil, their leaves able to capture light above the water's surface. Where the ground rises above the water but is saturated most of the time grow shrubs such as cranberry and some trees, including cedar and black spruce, that are adapted to wet soil. On nearby uplands, a foot or two higher, there may be white pines or white birches, which grow best in a dry, sandy soil.

Similarly, there can be many paths to an understanding of our place in nature. But like a traveler passing through a swamp who seeks a middle ground, neither walking waist-deep in the tea-colored water nor continually trying to jump from one dry hummock to the next, my goal is to seek a balance between the extremes of "civilization be damned" and "nature be damned."

There is a thin line, too, between the introspective, thoughtful attempt to find one's place in nature—Thoreau's approach—and the superficial touch-and-feel acquaintance of a passerby. To keep from crossing that line requires a certain appreciation of nature and a certain approach to knowledge, the topics of the chapters that follow. There is also a common misunderstanding of the different kinds of nature-knowledge, especially scientific knowledge and methods on the one hand and, on the other, beliefs based on intuition.

What hope, then, can those who love nature have for the human ability to sustain and conserve it? Henry David Thoreau appreciated, loved, revered, and took joy in both nature *and* civilization. He read the classics; he quoted poetry; he tried to write as best he could; he admired great men of the past who loved literature. He loved learning of all kinds. He sought both a spiritual and a physical connection between himself and nature. He was as concerned about the future of human civilization, as exemplified by his writings about civil disobedience, as he was about the future of nature. Throughout his life, he struggled to find a path to knowledge of many kinds and to understand his part in the relationship between people and land. His struggle is our struggle, and his clear thinking, his great ability to observe, can be of help to us. His life and story are encouraging to me, and I hope that as they are presented and interpreted on these pages, they will encourage others as well.

Daniel B. Botkin
Santa Barbara, California
August 15, 2000

NO MAN's GARDEN

Chapter 1

Climbing Mount Katahdin

"Who are we? Where are we?"
Henry David Thoreau,
near the summit of Mount Katahdin, Maine

"I looked down the dizzy abyss . . . at least a thousand feet deep, and walled up by perpendicular precipices. The scene was intensely sublime. The emotion was indeed overwhelming. On one side was the naked mountain peak, drear and desolate, its rocks rived by the frosts of six thousand winters. . . . The scene made the blood run chill and the teeth chatter."
William C. Larrabee, on his 1838 ascent of Mount Katahdin

On August 31, 1846, Henry David Thoreau left his home in Concord, Massachusetts, and went to the woods of Maine to discover wilderness for himself, rather than take other people's word for what it was like and what it meant. Direct, personal experience is the most obvious way to learn about nature and one's connection to it, and it was Thoreau's primary approach to knowledge. On this day, he sought to climb to the summit of Mount Katahdin to confront nature at its rawest. At 5,268 feet, Katahdin is the tallest mountain in Maine and the third highest east of the Rocky Mountains.

Thoreau arrived on the slopes of Katahdin in part because of his mentor, Ralph Waldo Emerson. Emerson was one of a group of intellectuals around Concord and Boston who had adopted the philosophical-religious doctrine of Transcendentalism. Among its precepts were the tenet that biological nature played a central role in religion and in human life in general, the belief that nature could be "read" spiritually, and the idea that studying God in nature could substitute for reading the Bible. Another precept was the belief that nature was benign and concerned about human beings. While his intellectual colleagues in this group were content sitting in their drawing rooms and discussing these beliefs—in fact, they considered insight superior to both experience and logic as a path to basic truths—Thoreau set off to explore their ideas, to test the truth for himself through firsthand observation. He hoped to achieve a spiritual connection with benign nature by experiencing it directly and by reading it as one would the Bible.

If you were to put a pin in a map at the geographic center of Maine, you would come close to the location of Mount Katahdin. It is part of the Appalachian Mountains—the backbone of eastern North America, extending some two thousand miles from Labrador to Georgia. The Appalachians are among the oldest mountains in the world, with some of their exposed bedrock more than 600 million years old—predating the Cambrian period, the time of the first great explosion in diversity of animals. Today, Mount Katahdin is the northern end of the Appalachian National Scenic Trail, which was developed independently by hikers in the 1920s and 1930s and made part of the National Trail System by the United States Congress in 1968. Katahdin is a massive mountain with several more or less isolated peaks rising over the rolling and flat terrain of the interior boreal forests of Maine.

On Sunday, September 7, 1846, Thoreau climbed partway up the mountain, reaching an elevation of 3,800 feet. "The wood was chiefly yellow birch, spruce, fir, mountain-ash, moose-wood. It was the worst kind of traveling; sometimes like the densest scrub-oak patches with us," he wrote, the "us" referring to residents of Concord, Massachusetts, and the surrounding countryside. Partway up, Thoreau left his companions to make camp while he continued ascending to find a route to the summit for the next day. Alone in the wilderness, he found himself "in a deep and narrow ravine, sloping up to the clouds, at an angle of nearly forty-five degrees, and hemmed in by walls of rock, which were at first covered with low trees, then with impenetrable thickets of scraggy birches and spruce-trees, and with moss." He was climbing through the stunted, gnarled, and twisted trunks of the last trees able to survive at those high altitudes, a shrub-

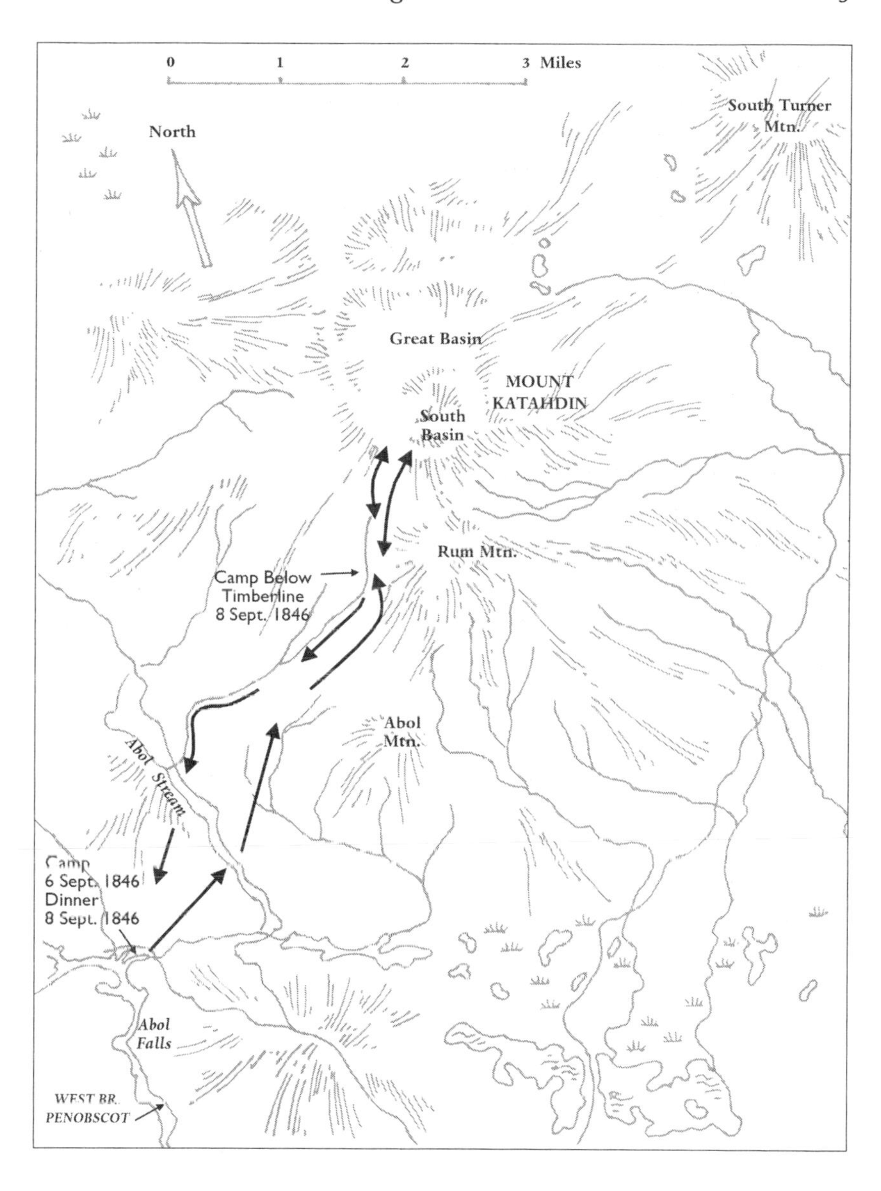

like growth called krummholz. He climbed "on all fours over the tops of ancient black spruce" and found "their tops flat and spreading, and their foliage blue, and nipped with cold, as if for centuries they had ceased growing upward against the bleak sky, the solid cold."

There were holes in the krummholz. Once, slumping through, Thoreau saw down "into a dark and cavernous region" and thought that the "holes were bears' dens, and the bears were even then at home." "This was the sort of garden I made my way over," he wrote, "certainly the most treacherous and porous country I ever traveled." It was seeming less and less like a benign environment.

Thoreau was one of the first people to make it this far up Mount Katahdin, no simple feat given the isolation of the peak and the limitations of the equipment available to him. The Indians believed that the mountain was too sacred to climb and rarely attempted it, and few Europeans had yet come this way with the curiosity and sense of adventure to climb the mountain. Through this difficult wilderness he struggled, pulling himself up "by the side of perpendicular falls of twenty or thirty feet, by the roots of firs and birches, and then, perhaps, walking a level rod or two in the thick stream . . . ascending by huge steps, as it were, a giant's stairway, down which a river flowed."

Briefly, he achieved what so many seek in such a climb today: a view of the countryside. "I had soon cleared the trees and paused on the successive shelves to look back over the country," he wrote, but the view was "almost continually draped in clouds."

After this partial ascent, Thoreau returned to his companions and the camp they had made, reaching them at dusk. There, the physical contact with nature began to affect his emotions. "When I returned to my companions, they had selected a camping-ground on the torrent's edge, and were resting on the ground; one was on the sick list, rolled in a blanket, on a damp shelf of rock. It was a savage and dreary scenery enough; so wildly rough, that they looked long to find a level and open space for the tent."

Viewing the camp, he wrote, "It was, perhaps, even a more grand and desolate place for a night's lodging than the summit would have been, being in the neighborhood of those wild trees, and of the torrent." This night, the evening before his attempt to reach the summit, he could still write of a desolate piece of wilderness as "grand." This was to change the next day, when Thoreau climbed to what he believed was the summit, reaching it considerably ahead of his companions. There, he "climbed alone over huge rocks, loosely poised, a mile or more." The terrain was "concealed by mist," even though the day was otherwise clear. There were so many boulders that it seemed "as if some time it had rained rocks, and they lay as they fell on the mountain sides, nowhere fairly at rest, but leaning on each other." These seemed to be "the raw materials of a planet dropped from an unseen quarry, which the vast chemistry of nature would anon work up, or work down, into the smiling and verdant plains and valleys of Earth. This was an undone extremity of the globe; as in lignite, we find coal in the process of formation."

Thoreau never quite reached the summit of Mount Katahdin; what he did reach was a tableland between two peaks, later named South and Baxter, the latter the higher and the actual summit of the mountain. The tableland is much like the summit in its rawness and exposure. There, Thoreau found that he was "deep within the hostile ranks of clouds." The clouds blew in and out. "It was, in fact, a cloud factory. . . . Occasionally, when the windy columns broke into me, I caught sight of a dark, damp crag to the right or left; the mist driving ceaselessly between it and me."

Thoreau was well read in the classics, and the mountaintop began to seem more like an ancient, pre-Christian saga than a replacement for the Bible. He wrote, melodramatically, "It reminded me of the creations of the old epic and dramatic poets, of Atlas, Vulcan, the Cyclops, and Prometheus," not of a benign, Christian God, and he felt estranged here from nature in a way he had not experienced before. "It was vast, Titanic, and such as man never inhabits." Standing within this awesome setting, he, a human being, seemed less significant as nature appeared more powerful. He felt "more lone than you can imagine." In the thinner air on the mountaintop, "vast, Titanic, inhuman nature" had him "at a disadvantage, caught him alone." He felt as if he had lost "some of his divine faculty," one of the very things he had sought to capture and understand on the mountain.

The setting took on the aspect of a theater in which Thoreau was a player and nature the star. He gave nature a speaking part:

> She seems to say sternly, Why came ye here before your time. This ground is not prepared for you. Is it not enough that I smile in the valleys? I have never made this soil for thy feet, this air for thy breathing, these rocks for thy neighbors. I cannot pity nor fondle thee here, but forever relentlessly drive thee hence to where I am kind. Why seek me where I have not called thee, and then complain because you find me but a stepmother? Shouldst thou freeze or starve, or shudder thy life away, here is no shrine, nor altar, nor any access to my ear.

On the summit of Katahdin, Thoreau discovered that wild nature was overpowering: "This was that Earth of which we have heard, made out of Chaos and Old Night. Here was no man's garden, but an unhandseled

globe. It was not lawn, nor pasture, nor mead, nor woodland, nor lea, nor arable, nor waste land."

So why *did* Thoreau climb the mountain? It is easy to imagine him wondering this very thing as he struggled in the clouds among the house-sized boulders, alongside steep cliffs, atop stiff and stunted black spruce. On his way down, he noticed "now and then some small bird of the sparrow family" that "would flit away before me, unable to command its course, like a fragment of the gray rock blown off by the wind." The bird, perhaps, was a metaphor for the way Thoreau felt about his own precarious situation. "Perhaps I most fully realized that this was primeval, untamed, and forever untamable nature, or whatever else men call it, while coming down this part of this mountain."

The immediate, specific reason why he had come to the mountain was to find nature at its essence, to discover a benign, caring nature, a "book of nature" that would set him spiritually at peace with nature and with himself. Within New England, Mount Katahdin, untrammeled by human beings, at that time beyond the reach of human-lit fires or air pollution, would have seemed the ideal place for such an encounter.

But what had ultimately brought Thoreau to the mountain was his lifelong quest to understand the connection between himself and nature in all its aspects, physical and spiritual, and to understand the relationships—the effects of one on the other—between nature and civilization, both of which he appreciated deeply, loved deeply. He came, perhaps, to find a permanent, even eternal, meaning for the relationship between a human being and nature and between civilization and nature.

His climb on the mountain was therefore driven by much more than what we today would call a concern with an environmental issue or with wilderness alone. It was a concern with the fundamentals of human existence, of the human situation—with the ultimate questions he was bold enough to write down, as naive as they may have sounded, reflecting on this trip up Katahdin: "Who are we? Where are we?"

But on Katahdin, Thoreau did not find what he had expected. Instead of a benign, caring nature, a substitute for the Bible that he could easily read, he discovered a wild nature unlike anything he had known before: an alien nature, anti-spiritual rather than inspirational, foreboding rather than enlightening, depressing rather than uplifting, rejecting rather than accepting, barren rather than fertile. His experience at the summit was not merely a bad day in a vacation outing; it was a cathartic experience in a lifelong search. The unexpected alienation he felt at the summit provided not an end but another beginning to his quest.

Today, one might think of a mountain climb either as recreation or as a confrontation with issues perceived to fall within a narrow scope of a political and ideological movement called environmentalism. The word *environmentalism* has grown to have a wide variety of meanings and is used loosely but rather narrowly. Even the dictionaries give it a narrow meaning. One definition in the tenth edition of *Merriam-Webster's Collegiate Dictionary* is "advocacy of the preservation or improvement of the natural environment; esp: the movement to control pollution." In his book *Green Backlash,* Andrew Rowell comments, "When I talk about the environmental movement, I mean someone who is fighting for ecological protection." Rowell defines "the mainstream environmental movement" as "the Group of ten," which includes "Defenders of Wildlife, Environmental Defense Fund, National Audubon Society, National Wildlife Federation, Natural Resources Defense Council, Friends of the Earth, Izaak Walton League, Sierra Club, Wilderness Society, and the World Wide Fund for Nature." These organizations, Rowell writes, "focus their agenda on changing legislation in Washington." From these statements, one would believe that environmentalism is no more than political advocacy for improvement or preservation of the environment or, more loosely, as the term is often used, sympathy for such advocacy. But environmentalism is involved with—is fundamentally integrated with—much deeper issues.

It has become popular to say that one is an environmentalist. But when former president George Bush calls himself an environmentalist, I take it to mean that he wants to position himself politically as in sympathy with those who want a good environment, whatever "good" might mean. When the majority of Americans tell pollsters that they consider themselves environmentalists, I take it to mean both that they are in favor of an improved environment rather than a declining environment and that they are in sympathy with those who work toward environmental improvement.

Polls conducted by the Charlton Research Company of Walnut Creek, California, suggest that most people perceive environmentalism as rather narrowly focused—concerned more with "brown" issues, such as pollution of air, water, and soil, than with "green" issues, such as protection of endangered species and biological diversity, or issues at the root of the human condition.

The term *environmentalism* therefore generally refers to a political movement that began in the twentieth century as a reaction to the negative environmental effects of technology, civilization, and a large human population. When viewed in this light, environmentalism does not appear to deal with the fundamentals of contact between humanity and nature, nor is it seen as a philosophical movement. When seen in this narrow way, environmentalism can suffer from lack of a firm foundation in a philosophical position, although such positions are often implicit. As often practiced and pursued, it does not necessarily depend on scientific understanding or necessarily involve the pursuit of scientific knowledge, although environmental advocates often use the findings of science as they understand them. When seen as a political and social movement, environmentalism therefore differs from a religious or spiritual movement, although an individual might become an environmentalist because of spiritual or religious feelings. Many people who are unfamiliar with the long history of humanity's concern with its relationship to nature believe that environmentalism arose with little connection to this concern.

But the connection between human beings and nature strikes much more deeply within a person than do social pressures to be for or against or to participate in a political movement, just as the summit of Mount Katahdin struck deeply into Thoreau. As long as people in Western civilization have written, they have written about the relationships between the individual and nature and between civilization and nature. Three great questions have occupied many philosophers, theologians, poets, and artists and, in modern times, many scientists: What is the character of nature undisturbed by human influence? What is the influence of nature on people? What is the influence of people on nature? These questions are linked in that they seek an understanding of how nature and humanity are by necessity connected and ought to be linked for each other's benefit.

The challenges we face today are as difficult and unclear to us as was the summit of Mount Katahdin to Thoreau on his climb. Can both civilization and nature flourish? Can we find a place for ourselves within nature that satisfies our spirits as well as our material needs and desires? These questions are the fundamental, underlying concerns of environmentalism and the concern of this book. In this sense, the book is not about environmentalism per se; it is about the human drama as it plays out on an environmental stage, with the term taken in its broadest meaning. In the present scene of this drama, nature and civilization are at risk from two extremes. One extreme is human overpopulation combined with an extension of the destructive aspects of technology. Together, these might overwhelm the diversity of life on Earth and the Earth's life-maintaining processes. The other extreme is the danger of civilization's collapse if we give way to a completely biocentric value system, believing that to save nature, we must sacrifice the lives of vast numbers of people and the processes and qualities that make civilization possible.

Thoreau's experiences and ideas are invaluable in our search to avoid either extreme because both civilization and nature mattered deeply to him. His approach to both civilization and nature is instructive.

We face a challenge that few, if any, peoples and civilizations before us have surmounted. Given a sufficient number of people and sufficient technology, most past civilizations have overused their resources, and by this means, some appear to have brought about their own downfall. Today, with the greatest numbers of people and most advanced technologies ever on Earth, we must find ways to use our human resources and technologies to sustain nature and civilization.

Solutions will not simply arise from without; they will depend on our inner perception of our connection with nature. If we do not care for nature, nature will suffer. If we do not experience nature, both we and nature will suffer. If we do not have an inner sense of our connectedness to nature, our quest for a path that will allow civilization and nature to persist and prevail will not succeed. Forging this path will require each of us to understand the way we perceive our connection with nature, both physically and spiritually. Here, I use *spiritually* in a broad sense to include a sense of wonder and awe at nature, a feeling of the grandeur and power of nature's intricate and wonderful systems of living creatures, species, communities, ecosystems—the Earth's biosphere. The term *spiritual* also includes, for those who receive it, a sense of something larger: the metaphysical, whatever one wants to call it, from an eminence of nature to God. These ideas are consistent with the responses of participants in a recent public opinion poll focus group. These people said that they went to nature for beauty, by which they meant a place that gave them an experience that took them out of their ordinary lives, uplifted them, and made them better persons.

We are at a critical juncture, for nature and for civilization. On the one hand, the scale of human presence and the power of our technologies enable us to alter the world fundamentally, for good or ill. On the other hand, the power of our technologies and the depth of our new understanding give us the capability to address these problems as never before.

Thoreau understood that these issues are intertwined. His search for a spiritual connection to nature drew him to the mountain and affected the way he approached it physically, and his responses to those physical experiences profoundly influenced his religious and spiritual beliefs. His perception of his place in nature affected his ideas about civilization and nature, and the effects of civilization on nature and of nature on civilization affected his perception of his own role and place in nature. Thoreau's climb up the mountain was a confrontation with himself and with civilization as well as with nature.

Thoreau devoted much of his life to exploring these fundamental issues. Although he saw these questions from the perspective of the early machine age and the beginnings of biological science, he saw them whole. Thoreau's writings about his travels to Maine and elsewhere in New England, as well as his walks in the woods near his home and his life in the town and city, provide keys to unlock new ways of thinking for us at the beginning of a new millennium.

In our search, we are as likely to discover things we do not know and do not expect, as did Thoreau on Mount Katahdin. We must face these challenges directly, and clearly, just as he did in ascending the mountain. The country ahead of us is undiscovered and may be no more comforting than a crawl over krummholz where a bear appears to lurk below, where the tableland is full of clouds, mist, and boulders. But we will be dragged there one way or another—and if not by conscious choice, so much the worse for us. Let us, then, attempt to ascend the mountain and, on our descent, seek a better understanding of nature, civilization, and ourselves; seek reflection as Thoreau did on his return from the Maine woods; and seek a variety of paths to knowledge, as he did throughout his travels and throughout his life.

Chapter 2

Crossing Umbazooksus Swamp

"I suspect that, if you should go to the end of the world,
you would find somebody there going farther, as if just starting for
home at sundown, and having a last word before he drove off."
Henry David Thoreau, *The Maine Woods*

"With exaltation at the beauty of nature comes wonderment as well,
and the belief that man comes closer to the heartbeats of the creation when
he is alone in primordial harmonies, away from other men and their artifacts,
because unlike the haunts of men these are sacred precincts."
Clarence J. Glacken, *Traces on the Rhodian Shore*

ELEVEN YEARS after his ascent of Mount Katahdin, Henry David Thoreau returned to the Maine woods for his third and last trip, canoeing and hiking for several weeks in the forests, lakes, and streams with his frequent traveling companion Edward Hoar. On July 27, 1857, rather than struggling up a mountain, Thoreau, at age forty, found himself lost in Umbazooksus Swamp, a large wetland deep in the Maine woods about 100 miles north of Bangor, Maine, and about 280 miles from his home in Concord, Massachusetts. Just to the south was Chesuncook Lake, frequented today by recreational canoeists. Thirty miles to the east rose Mount Katahdin, the nemesis of his earlier trip, obscured to him that day by the dense trees and shrubs typical of northern wetlands.

This trip was a travel back in time as much as to a distant place. Disembarking at Bangor from an oceangoing steamship, Thoreau traveled by wagon to Oldtown, a small village famous today for canoes. From there he traveled by bateau, the boat of the voyageurs who had explored North America's wilderness forests several generations earlier. The voyageurs had sought furs in forests like the one Thoreau was now visiting. Thoreau took the bateau to an island where Penobscot Indians lived. There he hired as his guide Joseph Polis, a Penobscot Indian who was well known in the region as an expert on the woods, nature lore, and Indian culture.

It is curious to think of Thoreau, one of the fathers of modern environmentalism and author of the famous aphorism "In wildness is the preservation of the world," as ever having been lost in wilderness. He is best known for his long essay *Walden,* based on his two-year sojourn in a cabin he built by himself in the woods near Walden Pond in Concord, an essay that is often required reading in high school or college. From audiences I have spoken before and in conversations with many people around the United States, I have found that few know much more about him, even though he is a central historical figure to those involved in the environmental movement or interested in the history of conservation. Most people with whom I have spoken are unaware of Thoreau's other books—*The Maine Woods, Cape Cod, A Week on the Concord and Merrimack Rivers*—or his many essays about nature and his extensive journals.

The image of Thoreau cast by *Walden* is of an ill-kept, bearded man living a hermit-like existence, alone in the woods contemplating nature, with little interest in other people. Wilderness would seem to be the perfect place for him. But on this day in 1857, there he was, as he described himself later in *The Maine Woods,* sitting on a raised hummock in the midst of a large swamp—lost.

Thoreau was not only lost in a physical, geographic sense; he was lost among his ideas and feelings, seeking in those dark woods and vast wetlands a spiritual as well as physical connection with nature. He was lost in thought about his surroundings—wondering what the squirrels and birds were thinking and saying to themselves about the woods. Over his lifetime, Thoreau wrote much about his ideas concerning nature and people, sometimes contradicting himself, changing his mind and then going back to an earlier idea—not always certain of what he believed. Eleven years after his climb up Mount Katahdin, he was still searching for ultimate answers.

Thoreau, Polis, and Hoar had canoed into the divide between the two major watersheds of northern Maine marked by Umbazooksus and Mud Lakes, trying to find their way to Chamberlain Lake in the north. Their eighteen-foot canoe was loaded down with two large knapsacks as well as two large india rubber bags, which Thoreau had brought for his gear, and their guide's ax, gun, pipe, tobacco, and blanket, "all the baggage he had." Thoreau estimated that the group's baggage weighed 166 pounds.

At the divide, the waterway had become too shallow to carry three people and this amount of gear. Polis had told them that he would take the canoe around the swamp while Thoreau and Hoar walked through it, carrying some of their gear across the wetlands. Thoreau and Hoar were thus on a "carry" or "portage"—a standard part of canoe travel through lake and stream country. They were to meet Polis at Chamberlain Lake.

They were in the midst of the great boreal forest, in what Thoreau called "the wildest country," a forest of spruce, hemlock, cedar, and fir, white and yellow birch, white pine, and aspen. It is one of the major forests of the world, stretching outward from Umbazooksus Swamp to the borders of northern and western Maine, into New Brunswick and north to the timberline in Canada, almost to Hudson Bay; south to southern New Hampshire and the hilltops of Appalachia; west to the prairies of Minnesota and Saskatchewan; and farther north and west still, reaching around the world at high latitudes into Siberia, northwestern Europe, Finland, Sweden, and Norway. The boreal forest remains one of the major forests of the world, covering more than 4 million square miles, about 6 percent of the total land area of the planet, and still containing some of the wildest country on Earth.

In the wetlands of these forests, spruce and fir form dense, thick stands, creating a dark forest often difficult to traverse, where one pushes through needle-clad branches and over dead logs only to find that the ground has fallen away into a still wetter forest of larch trees or, worse, into a bog of sphagnum moss and shrubs and patches of open water. Distant vistas of a grand panorama are rare in boreal forests, restricted to clearings and ridge tops.

Joe Polis had given Thoreau and Ed Hoar explicit directions, from his point of view, about how to cross the wetlands without getting lost. He told them simply to follow tracks he had made previously, assuming that any self-respecting woodsman could track another.

Thoreau and Hoar had divided the gear and set off directly, although at

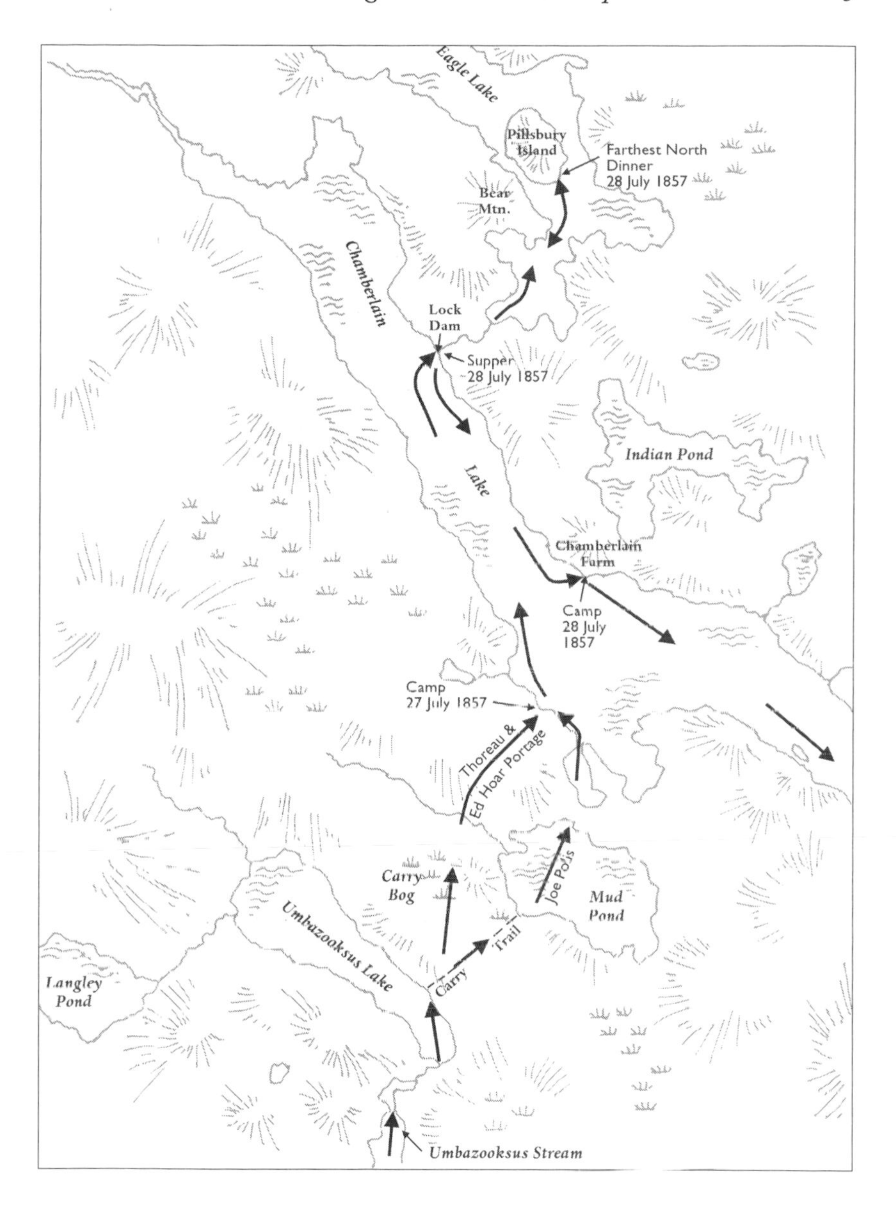

the outset, Thoreau had been skeptical about the success of their woodland navigation through the portage. "I had not much faith that we could distinguish his tracks," he later wrote, "since others had passed over the carry within a few days." As he expected, they were soon lost.

At first, the two wilderness novices found themselves in a cedar forest, "an arbor-vitae wilderness of the grimmest character," Thoreau noted, where "it was impossible for us to discern the Indian's trail in the elastic moss, which, like a thick carpet, covered every rock and fallen tree, as well as the Earth."

Thoreau's load was heavy. He estimated that he carried about sixty pounds of equipment in a knapsack and one of his large india rubber bags; on this day, the rubber bag held bread, bacon, utensils, and a blanket—good things to keep from falling into the tea-colored waters. Through much of Umbazooksus Swamp, the men sank a foot or more, sometimes up to their knees, into water and mud.

Whereas Thoreau carried his portion of the gear all at once, Hoar divided his burden in two, moving one part ahead and then going back for the rest. Each time he did this, Thoreau rested and observed the woods. Blackflies swarmed about him. He tried a "wash" made of turpentine, spearmint oil, and camphor as a repellent, but after applying it to his face decided that "the remedy was worse than the disease."

In spite of the insects, Thoreau was able to use the time to observe and reflect on the woods and its inhabitants. Three Canada jays came by. Fish hawks whistled above the lake. A white-throated sparrow called. A red squirrel caught his attention. "It must have been a solitary time in that dark evergreen forest, where there is so little life, seventy-five miles from a road as we had come," Thoreau wrote of the squirrel, expressing his own sense of aloneness and separation in the wild woods. "I wondered how he could call any particular tree there his home; and yet he would run up the stem of one out of the myriads, as if it were an old road to him. How can a hawk ever find him there? I fancied he must be glad to see us, though he did seem to chide us." This was as much as to ask, What do the squirrel and the hawk know about wilderness that I do not, so that they can call it home, feel it to be home, know it so well that the tangles of indeterminate branches are to them a friendly neighborhood? Why could Thoreau not capture that same feeling of oneness with nature, an at-home-ness in this wettest and wildest of all lands that he knew? Why could he not find his way through it as well as did the squirrel or the hawk or Joe Polis, his guide?

Thoreau was rejoined by Ed Hoar, and the two continued their wetland struggle. "The walking was worse than ever," Thoreau wrote, because of fallen timber as well as wet ground: "The fallen trees were so numerous that for long distances the route was through a succession of small yards, where we climbed over fences as high as our heads, down into water often up to our knees, and then over another fence into a second yard." The time Thoreau had between his companion's carries to ponder questions about human beings and nature was not long enough for him to arrive at answers, but the chattering squirrel seemed to know the connection.

The carry was taking all day, and as Thoreau waited again for his companion to retrieve his equipment, he realized that the sun was getting low. Being lost for a night where there was no dry ground to lay a blanket or to light a fire whose smoke at least would discourage insects was not a happy prospect. As the day drew to its close, the two decided to push onward through the final leg without Ed Hoar going back one last time for the rest of his burden. They hoped instead to find Joe Polis before dark and retrieve their materials the next day.

About a mile farther, they "heard a noise like an owl, which I soon discovered to be made by the Indian, and answering him, we soon came together," Thoreau wrote. "If he had not come back to meet us, we probably should not have found him that night."

Joe Polis found Thoreau and Hoar only because he had gone back to the camp of a Canadian and asked him which way they had probably gone, "since he could better understand the ways of white men." Polis was "greatly surprised" at the way they had taken. He "said it was 'strange' and evidentially thought little of our woodcraft," Thoreau wrote. Thoreau still had much to learn about the ways of nature and the methods for finding his way through it.

Polis was an important figure for Thoreau, both a spiritual and a physical guide. His knowledge of the woods impressed and amazed the Concord gentleman. Thoreau admired Polis's woodsmanship and skills in wildland navigation. What was lacking in him that was within Joe Polis? Was it a lifetime of careful natural history observation—careful observation of the tangible reality of his surroundings? Or was it some inner, spiritual sensitivity developed by being within nature rather than by consciously observing and thinking about it? Sitting on the hummock in Umbazooksus Swamp, Thoreau had not been sure.

It was a question he had been asking himself throughout the trip. Two days before, on July 25, when they were northeast of Mount Kineo, alongside Moosehead Lake, Polis had told Thoreau that he made most of his money as a hunting guide. Thoreau asked Polis how he found his way through the woods. Polis told him that he looked at hillsides because there was a "great difference between the north and south" in the trees. Polis also said that he looked at the rocks, but he did not, or could not, explain to Thoreau what it was about the rocks that told him the compass directions.

Then Thoreau asked him, "[If] I should take you in a dark night, right up here into the middle of the woods a hundred miles, set you down, and turn you round quickly twenty times, could you steer straight to Oldtown?" Polis replied that he could and had done "pretty much the same thing." He told Thoreau he had once taken an experienced white hunter into the woods to shoot a moose. They had found a moose, but it had led them "round and round" most of the day. Polis had then asked the hunter, who had said previously he could find his way anywhere in the woods, to take them directly back to Oldtown. The hunter said he did not know where he was and therefore could not lead them back, and so Polis led the way. "Great difference between me and white man," Polis told Thoreau.

To Thoreau, Polis seemed to use so many kinds of information "that he did not give a distinct, conscious attention to any one, and so could not readily refer to any." Instead, he seemed to find his way intuitively, much as did the squirrel or hawk. Perhaps, Thoreau thought, in some way it was "merely a sharpened and educated sense." Or was it more? Was the path to understanding nature through observation of the tangible, physical phenomena, through a spiritual connection with nature, or both?

Although the story of one of the first environmentalists becoming lost in a swamp and needing the aid of his Indian guide may be curious, it has a deeper importance. Modern science and technology are changing the way we understand and find our way through physical nature and therefore are changing our physical relationship with it and our perception of it. Hikers today carry satellite-guided global positioning systems and cellular telephones; some lost hikers have called the police and asked that helicopters come and rescue them. Satellite images provide new kinds of maps that give us a new perspective on forests. The science of ecology and related environmental sciences are changing our understanding of how nature works and how the nature we observe came to be. The dominant scientific idea about nature during the nineteenth century and most of the twentieth century was a continuation of the ancient idea in Western civilization of the balance of nature, an idea that can be traced back to the Greeks and Romans and beyond. This is the idea of nature as constant and perfect in its constancy; of nature in a steady state; of a nature that if left to itself would achieve a permanence of form and substance that, unless disturbed, would continue indefinitely.

Scientific research in the last quarter of the twentieth century toppled that notion. Modern ecological science tells us that nature changes at every scale of space and time, that there are many kinds of natural change, natural in rate and amount. Life has evolved and adapted to these changes for 3.5 billion years. Take away many kinds of change, or greatly alter their rate and strength, and many species, if not all, will be maladapted to the new, static environment, and some or many may go extinct. In the Maine woods, for example, disturbance and change are necessary for white pine, white and yellow birch, and aspen. White pine, one of the most valuable of the northern trees and the first to be cut in the Maine forests, regenerates only where ample light reaches the ground, such as occurs after a storm or fire. White birch seeds, which are small and light, float and move easily in the wind and travel long distances. But these seeds must fall on bare, mineral-rich soil to germinate and survive. Seedlings and saplings of white birch also require bright sunlight near the ground. This species is adapted to clearings created by fire or storms sufficient to turn over the soil or scrape it clear of leaves, twigs, and organic mulch. The woods through which Thoreau walked represented the adaptation of life to a great many kinds of change over time.

Other species, such as spruce and fir, which formed the dense stands that were so hard for Thoreau to walk through, grow well in shade and tend to dominate older stands that have not been disturbed recently. Over time, spruce shades out and replaces white pine and birch on the drier sites. Each species has its place in the unfolding story of the growth of a forest after a clearing. It is a story that Thoreau was one of the first modern naturalists and scientists to understand.

It is a story that affects humanity. The new, radically different understanding of nature creates a perspective that can be uncomfortable, at least initially. To abandon a long-standing cultural myth is always difficult, but to replace a myth of constancy with an idea of change is especially disconcerting. Most people I know like permanence. They like to know that the places they visited years ago will look the same when they return. The new scientific findings about nature tell us that often this will not be the case, that landscapes we love can change dramatically over time, and the changes are natural. How can we deal with this? How can we learn to live and be comfortable with a nature that is always changing? What are the implications for our physical and spiritual connections to nature and, therefore, our entire sense of ourselves, our role, our purpose in life? Where do we fit into nature? And in terms of our physical needs and desires, what are the implications for the way we conserve, manage, and use nature's resources, its forests, wildlife, fish, and soils? What are the implications of this new idea for the place of wilderness in our landscape and the kinds of wilderness that will be there? These are questions Thoreau explored throughout his life. His ideas can be helpful to us.

The symbolism of a swamp remained with Thoreau, and he returned to it often in his lectures and his writing. Thoreau was a surveyor and often made his living that way. In one of his mature essays, "Walking," he described a survey of a swamp that he conducted for a farmer near Concord. The swamp was so deep that he saw the farmer "up to his neck and swimming for his life in his property, though it was still winter." The same farmer had "another similar swamp which I could not survey at all, because it was completely under water," and a third "which I did *survey* from a distance," Thoreau wrote, and about which the farmer said "he would not part with it for any consideration, on account of the mud it contained." This might be interpreted as simply an expression of a physical utility of the swamp, but Thoreau used it here, as he did frequently, to symbolize the deeper meanings of swamps for himself. Even a swamp that would overwhelm a farmer was too valuable to give up. So it was with a person's relation to nature.

Thoreau made it out of Umbazooksus Swamp thanks to the woodland knowledge of Joe Polis and Polis's quick thinking about how he might find the two lost travelers. But Thoreau still did not resolve the issues that concerned him as he sat on the wetland hummock wondering about the place of the birds, the squirrel, and himself within nature. This resolution seemed to require more exploration. Not only did the image of the swamp remain with Thoreau throughout his life; the metaphor of being mired in Umbazooksus expressed a tension that continued for the rest of his life and drove him to seek yet more knowledge of nature.

Chapter 3

Enjoying the Swamp on the Edge of Town

"In Princeton College, they had talked a good deal about civilization. . . . Up to that time, civilization had just been a fancy word that preachers and professors and politicians bruited about."
Larry McMurtry, *Streets of Laredo*

Where do the Blues come from?
Come back in the lowlands, where my folks are from.
Traditional blues song

EVEN THOUGH the going had been tough and in many ways unpleasant in Umbazooksus Swamp, the place had deeply impressed Henry David Thoreau, and the image of a swamp remained with him throughout his life. In contrast, though the going had been tough during Thoreau's climb up Mount Katahdin and the effect of that ascent also profound, the image of a mountain did not persist with him as a central theme or metaphor. It was the biologically rich swamp, not the barren geological formation of the mountain, that Thoreau retained as a lasting impression and image.

His comments about swamps give us insight into his general perspective on nature and civilization. Later in life, Thoreau wrote, "Yes, though you may think me perverse, if it were proposed to me to dwell in the neighborhood of the most beautiful garden ever human art contrived, or else of a Dismal Swamp, I should certainly decide for the swamp." This assertion would seem to be a complete rejection of civilization: a decision to sacrifice civilization to preserve nature, perhaps, or a desire to live in uncivilized nature, consistent with the ideology of the most extreme environmentalists of our day.

But this was not the case. For Thoreau, a love of swamps was not a personal commitment to life solely within the depths of wilderness. At the end of his second trip to the Maine woods, he wrote, "It was a relief to get back to our smooth, but still varied landscape." The place he chose to live was within civilization: "It seemed to me that there could be no comparison between this [the village of Concord and its surroundings] and the wilderness, necessary as the latter is for a resource and background, the raw material of all our civilization." Thoreau did not reject civilization for nature or nature for civilization. He believed instead that one function of wilderness—and therefore of all nature—was to benefit people and civilization. He did not value wilderness or nature simply for itself.

The idea of wilderness that was dominant at the end of the twentieth century—as a place untrammeled by human beings, where there is no evidence of the action of people—tends to contrast civilization and wilderness as the white and black, or black and white, of nature and to lead to a landscape that can seem to be only one or the other. But our need is to discover what overall mixture of colors on the landscape works best for both nature and civilization—what mixture sustains both of them and the material and spiritual values of greatest meaning to people. Thoreau himself enjoyed a range of landscapes, as made clear by his experiences described later in this book.

But at this stage in his life, having had his fill of the big woods, Thoreau found that "the wilderness is simple, almost to barrenness." What he really liked, then, was a combination of civilized—settled—countryside and some access to wilderness. "The partially cultivated country it is which chiefly has inspired, and will continue to inspire, the strains of poets, such as compose the mass of any literature. Our woods are sylvan, and their inhabitants woodsmen and rustics," he wrote. "Perhaps our own woods and fields . . . are the perfection of parks and groves, gardens, arbors, paths, vistas, and landscapes." He liked a swamp on the edge of town.

The swamp, he believed, was important to civilization: "A town is saved, not more by the righteous men in it than by the woods and swamps that surround it. A township where one primitive forest waves above while

another primitive forest rots below,—such a town is fitted to raise not only corn and potatoes, but poets and philosophers for the coming ages."

There was no doubt in Thoreau's mind about the importance of nature to civilization. Later in life, he wrote: "The civilized nations—Greece, Rome, England—have been sustained by the primitive forests which anciently rotted where they stand. They survive as long as the soil is not exhausted. Alas for human culture! Little is to be expected of a nation, when the vegetable mould is exhausted, and it is compelled to make manure of the bones of its fathers. There the poet sustains himself merely by his own superfluous fat, and the philosopher comes down on his marrow-bones." Nature was necessary for the best civilization—that was one of nature's functions—but nature without civilization was not a place for Thoreau. Nature without civilization had no meaning for human beings.

This is an important perspective, given that late-twentieth-century Western civilization has made wilderness into a strongly positive value. Few, if any, civilizations have made this judgment, instead perceiving wilderness as something to be feared, something ugly, or, at best, something fearsome but sacred.

The way in which one views the relative value of civilization and nature—nature, in this case, best revealed in wilderness—influences one's perception of what each should be. In Thoreau's time, the latest technology—the result of the flowering of machine-age civilization—was just beginning to make vacation trips into the wilderness possible. Thoreau made use of some of this technology to reach Umbazooksus Swamp. He traveled from civilized Massachusetts to civilized Bangor, Maine, by steamer, which had been invented in 1807, ten years before his birth, and had come into great commercial importance only during his lifetime. He carried much of his gear in two large india rubber bags, of which he was especially proud as the latest in backcountry technology, equivalent in his time to the latest carbon fiber Gore-Tex backpack of wilderness seekers today. At the very least, one can say that the technology of the time enhanced the ease with which Thoreau could reach the Maine woods and therefore made his contact with this wilderness possible within the rest of his activities.

Thoreau's india rubber bags epitomized some of the physical aspects of contact between nature and civilization. A product of *Hevea brasiliensis,* a tree of the Amazon Basin and other Central and South American rain forests, rubber had been known to the Indians of those regions since at least the eleventh century. As part of European exploration of the New World, Charles de La Condamine, a Frenchman interested in natural history, sent the first samples to Europe around 1740. The great early chemist Joseph Priestley examined the properties of this strange substance and gave it its name. But it was not until 1818 that a major practical use for it was found. A British medical student named James Syme used rubber to waterproof a raincoat, later to become known as the mackintosh. And it was not until 1839, when Thoreau was twenty-two, that Charles Goodyear developed the vulcanization process, a mixing of rubber and sulfur, that made commercial products such as Thoreau's bag practical. A product of a tree from one wilderness, a tropical rain forest, was transported a quarter of the way around the world to France and England, studied by scientists, experimented with by inventors, and developed into commercial products that were shipped to another part of the world, Thoreau's Massachusetts. There, it aided human contact with and alteration of another wilderness, the Maine woods of Thoreau's time. This was, for Thoreau, a positive effect of the civilization side of the civilization–nature relationship—it got him into the Maine woods. Civilization's technology—and therefore its creativity—helped Thoreau make contact with nature and thereby contemplate his connection, both physical and spiritual, with nature.

And Thoreau greatly loved and appreciated civilization. In 1839, he and his older brother, John, built a boat "in form like a fisherman's dory, fifteen feet long by three and a half in breadth at the widest part." They painted it green below with a blue border, made wheels for portaging it around falls and dams, and provided it with two masts, two oars, and two poles for maneuvering in places too narrow to row. Thoreau wrote that the boat "cost us a week's labor in the spring."

On Saturday, August 31, 1839, Henry David and John pushed their boat into the water and set off on a trip along the Concord and Merrimack Rivers. Thoreau had graduated from Harvard University only two years earlier, and he was unsure of his future career and just beginning to formulate his ideas. He kept a journal of this trip, as was his custom throughout his life.

Three years later, in 1842, his brother died. While living at Walden Pond, Thoreau worked on a manuscript based on their boating trip. He completed the book in 1845 and, unable to find a publisher, published it at his own expense. Some say that Thoreau wrote this first book, *A Week on the Concord and Merrimack Rivers,* as a tribute and memorial to his brother. It was not a success at the time. Having paid for its publication—1,000 copies—he sold only 215, gave away 75, and kept the rest. He joked that he had a library of hundreds of books, most of which were his own—the unsold copies of this first book.

Thoreau's commentary about his journey on the two rivers begins simply enough, with a description of nature's beauty. As he and John were leaving Concord, Thoreau wrote, "gradually the village murmur subsided, and we seemed to be embarked on the placid current of our dreams, floating from past to future as silently as one awakes to fresh morning or evening thoughts." It was romantic, as one expects books about nature to be. "Nature seemed to have adorned herself for our departure with a profusion of fringes and curls, mingled with the bright tints of flowers, reflected in the water." Thoreau's book began as did many later romantic books about nature and many nature documentary films of the twentieth century: observing a pretty scene, perhaps with nothing deeper to think about than such scenery.

But the book soon digresses from a romantic description of tangible qualities of the countryside, from the viewpoint of an urban person two years out of Harvard, to a discussion of literature and other products of civilization.

It is easy to see why the book sold poorly. It rambles and seems unfocused, and its goals and approach were unusual. It describes a trip through the countryside as a backdrop for a discussion of issues profoundly important to Thoreau: philosophy, literature, civilization, and nature. Travel books were a novelty at the time, and few went beyond superficial description; they were guides to the tangible. Books of philosophy tended to be didactic and theoretical. *A Week on the Concord and Merrimack Rivers* fit neither of these categories. It was a philosophical book set into a travelogue. Because it appeared to be a travel book, scholars who did not know Thoreau were not likely to buy it. People seeking travel guides were not likely to purchase it once they had opened it and read some of its rather rambling philosophical passages. I can imagine them thinking, "What kind of travel guide is this, anyway?" And who at that time would have sought a travel guide to the relatively settled and well known region along the Concord and Merrimack Rivers?

Thoreau was inventing a new genre—nature writing that seeks understanding beyond description—and his first experiment was not a great success. This genre would achieve success with *Walden,* a book that also mixes a concern with nature with concerns for many other aspects of human existence. However, *A Week on the Concord and Merrimack Rivers,* for all its limitations as literature, explains important aspects of Thoreau's approach to knowledge: the use of civilization's written word and the importance to him of human culture and creativity.

Thoreau considered the creation of literature to be an innate and natural activity of human beings: "As naturally as the oak bears an acorn, and the vine a gourd, man bears a poem, either spoken or done . . . history is but a prose narrative of poetic deeds. What else have the Hindoos, the Persians, the Babylonians, the Egyptians done, that can be told? . . . The poet sings how the blood flows in his veins." Thoreau was out in nature and, in part, trying to discover nature, but his focus was on human beings and civilization.

In our time, being out in nature is often seen as taking a break from book work and paperwork, and many people in the professions of natural resource management are those who prefer outdoor activities to the study of the classics. But for Thoreau, learning from the ancients and learning from his own observations of nature were intertwined. Both required exertion, and both were worthwhile: "To read well, that is, to read true books in a true spirit, is a noble exercise, and one that will task the reader more than any exercise which the customs of the day esteem." Reading well required a training "such as the athletes underwent," he wrote. "No wonder that Alexander carried the Iliad with him on his expeditions in a precious casket. A written word is the choicest of relics."

Thoreau understood that the search for wisdom was hard work. The pursuit of knowledge was a skill that required practice. His interest in achieving wisdom by reading what great people had written is another example of the central role that civilization played for him. It was never nature alone but rather nature and human wisdom, expressed in books, that attracted him. While at Walden Pond, he wrote, "My residence was more favorable, not only to thought, but to serious reading, than a university." The classics especially occupied him. "I kept Homer's Iliad on my table through the summer," he wrote. "For what are the classics but the noblest recorded thoughts of man?" The task of understanding great writing is like the task of understanding nature. One who is not a reader of the classics, he continued, "might as well omit to study nature because she is old." His attempt to understand nature was therefore made within the context of the study of literature and within the culture and civilization of his time.

When he could not learn about nature directly and could not find an expert with whom he could speak directly, Thoreau sought to learn what the great thinkers, writers, and poets of the past and of his time had to say. *A Week on the Concord and Merrimack Rivers* was an early example of his use of great literature, including contemporary science, in learning about nature, a process he would use throughout his life.

Thoreau, then, was well acquainted with both classical literature and the writings of his contemporaries. In his discussion of the depth of Walden Pond, he wrote, "This is a remarkable depth for so small an area," and then commented: "Not an inch of it can be spared by the imagination. What if all ponds were shallow? Would it not react on the minds of men? I am thankful that this pond was made deep and pure for a symbol." He was aware of the implications of such measurements on intangible aspects of human contact with nature. Scientific knowledge, a product of civilization, was capable of affecting the human imagination beyond science itself.

On the other hand, Thoreau lived when civilization of the machine age was exploiting nature in the Maine woods and in many remote areas of the world by harvesting living resources—extracting them much as gold and silver are mined. In Thoreau's New England, forests were often seen as an impediment to progress and simply removed to make way for farms and towns. There was so much timber that often the timber was not even used. One common method of clearing land was to girdle most or all of the trees, cutting through the bark in a thin line all around the trunk. This prevents water and minerals from getting up to the leaves and prevents sugars from getting from the leaves down to the roots. It is the simplest way to kill a tree, easily done with an ax or saw, much easier than cutting the tree down. Once the trees were girdled, fires were set to clear the land and open it to settlement. In southern New Hampshire, today within commuting distance of Boston and Concord by automobile, these intentionally set land-clearing fires had so much fuel from the dead and dying trees that the fires burned hot and ranged beyond the intended areas.

One such fire burned to the very top of Mount Monadnock in New Hampshire, not only killing the trees but also burning away the organic soil; today, this popular hiking destination is bare rock at the summit, providing a beautiful view of the surrounding, now heavily wooded countryside. Monadnock's summit is unlikely ever to be forested again, short of the arrival of another ice age that might deposit sand, silt, and clay there.

Such actions suggest that those who cleared the land believed that nature had the upper hand and that it was a necessity—perhaps even an obligation—to correct the balance in favor of civilization. The society of Thoreau's time was decreasing the area that today would be considered "natural" or "wilderness" because it interfered with settlement, progress, and people's ability to make a living.

From a modern ecological perspective, this method of land clearing left an abnormal amount of fuel, and the resulting fires were, as a result, abnormally destructive. It was the kind of human-induced event that led to a belief that all forest fires were bad and unnatural.

Thoreau's discussions could be taken to have been much on nature's side, at least in terms of the way nature and civilization were perceived in his time. His appreciation of swamps and their deep organic material might be interpreted in part as a reaction against such misuse of the land and ignorance of the value of its material qualities, including the rich soils of many old forests. It might be interpreted as a reaction against the willingness of many New Englanders to wantonly destroy forests and soils as part of land clearing. Few during Thoreau's time wrote from such a perspective. It was not until two years after his death that his contemporary New Englander George Perkins Marsh wrote the first great modern book about the effects of civilization on nature, *Man and Nature.* Both Thoreau and Marsh appreciated and enjoyed civilization; each sought an appropriate combination and relationship of the two.

Today, it is easy to understand why some people have been so frustrated with civilization's effects on nature. The bases for the belief that modern civilization and nature cannot coexist are well known: our destruction of resources during the period of exploitation and our failure to sustain most living resources during the period of professional management to maximize yield of single resources. Society has seemed to extend modern civilization only at the expense of nature. Added to this are the continuing growth of the human population and the growing powers of science-based technology. These have included the development of the atomic bomb and other modern weapons, which seem capable of destroying most, if not all, life on Earth and causing the pollution of the air, water, and soil. Even artificial chemicals that initially appeared benign have had inadvertent environmental effects. The adverse effect of dichlorodiphenyltrichloroethane (DDT) on the reproduction of many species of birds was a landmark discovery of this kind. With this history, innovative technologies made possible by creative science appear likely to overwhelm nature and the Earth's life-supporting and life-containing systems.

The sprawl of our cities and suburbs and the huge numbers of people in them appear capable of overwhelming nature directly. Cities, the traditional centers of civilization, are growing in area and numbers of inhabitants. In the United States, more than 80 percent of the people are urban or suburban, highly dependent on modern technologies for production and transportation of all life's necessities as well as employment and recreation. Worldwide, there is an urbanization momentum: the human population is becoming increasingly concentrated. In undeveloped and developing countries, more and more of the citizens are moving to a country's single largest city. In the future in many of these countries, more than half the population will live in the single largest city. Modern civilization seems poised to engulf nature with the global equivalent of girdling trees and spreading destructively hot wildfires that consume even the most remote wild places.

Civilization now seems overwhelmingly powerful over nature and completely separate from it. Epitomized by the city, civilization in this way is especially dominant over pure nature—wilderness. To be in a city is to be out of nature; to be in wilderness is to be out of civilization, global positioning systems, digital watches, and Gore-Tex clothing aside. Civilization appears as fixed and solid—as permanent and unyielding—as cement on a city sidewalk, as the steel in a skyscraper. The hustle and bustle of city life, with pedestrians and automobiles scurrying about, reinforces the image of civilization as powerful in energy and activities, not merely in structures—an unstoppable force that is overtaking nature.

Meanwhile, many have lost faith in the ability of science and technology to succeed as a basis for conserving and managing nature. If these have failed in the past, what faith can we have in their future? And these are among the triumphs of twentieth-century civilization. What hope, then, can those who love nature have for human abilities ever to sustain and conserve nature?

A partial answer, as we begin the twenty-first century, lies with new ideas about nature that are emerging as a result of twentieth-century creativity, innovation, and technological development—products of civilization that may help us understand nature and our connection with it. Our technologies give us not only tremendous power over nature but also an incredible view of Umbazooksus Swamp, of the Maine woods, of the entire Earth. What do we see about life and civilization from these viewpoints? From a low-flying helicopter, it is difficult at altitudes much above 2,000 feet to see an individual in a large swamp. One sees vegetation and streams. Higher up, from 20,000 feet, an individual in the woods is not perceptible to the naked eye. From a satellite circling several hundred miles above the Earth's surface, life appears as a thin green layer across continents and a patchy green coating of algae and photosynthetic bacteria on the ocean's surface, mainly near the continents. Civilization is even less apparent at this altitude, revealing itself most clearly at night as bright centers of light radiating from major cities like yellowish spiderwebs. During the day, only careful inspection of the Earth's surface reveals a meager few human artifacts visible from space.

So, from space, life on Earth appears as a thin layer on the surface. Within this, invisible, are the human processes of cultures and civilizations: Joe Polis's knowledge of how to find his way through the forest, Thoreau's thoughts about human beings and nature. They seem fragile from this viewpoint. Human ideas, inventiveness, and creativity, the ability to work wood and metal and rock, are fragile, delicate, perhaps fleeting. Yet they are the source of the seemingly solid city streets. From whence do they come? An answer Thoreau gives is that this creativity comes from the juxtaposition of the deep soils of a swamp—the essence of wilderness's best—and a humanized, settled landscape. Without both of these, civilizations decline and die, reverting to wilderness. Civilization, whatever its seeming sustainability, whatever its present power, is in a sense delicate and certainly unusual in the universe. The writings of Thoreau and his literary descendants, scientist Rachel Carson, forester Aldo Leopold, and naturalist Peter Matthiessen, the works of the great landscape painters Albert Bierstadt and Frederic Edwin Church, Thomas Moran and Georgia O'Keeffe, arose not from civilization or nature alone but from a combination of the two. What best leads to such creativity? Is it a landscape in which people are rare and do not live in civilized groups? Is it a suburbanized world with all the human population distributed more or less evenly across the landscape so that we have neither cities nor wilderness? Is it a world in which we reside in cities and are allowed to visit wilderness but not to touch it?

All these seem somehow not the right design. That is because they arise from a view of civilization and nature as two separate systems. Civilization and nature are one system, not two. Seeing them as one leads us to appreciate the swamp on the edge of town, the deep organic layer beneath the city. That swamp is Thoreau's metaphor for the integration of the two, his insight into what is at the root of creativity and innovation. Human creativity could not exist without civilization, but also it could not prevail without contact between civilization and nature. Maintaining a culture that provides for both the material and the spiritual needs and desires of people requires that combination.

Chapter 4

Nurse Trees and Nature

"Others do not see what I see. I don't see what they see.
That's a matter of perception and now I simply say: 'Fine.' . . .
I get into a dialogue with nature and put the question to nature,
not to my colleagues, because that's from whence the answer
must come."

Jonas Salk

"The jay is one of the most useful agents in the economy of nature,
for disseminating forest trees."

William Bartram, nineteenth-century botanical explorer

AT HOME in Concord, Massachusetts, Henry David Thoreau kept up with the latest discoveries and controversies in the biological sciences, contrary to the stereotype of him as a lonely hermit avoiding civilization. He read Charles Darwin's *On the Origin of Species* in 1860, within a year of its publication. At that time, an important biological controversy concerned whether life could arise spontaneously, without birth, eggs, seeds, or spores from a parental generation. Although the notion of spontaneous generation seems silly today, it was taken seriously by some major figures in the early nineteenth century. Already involved in observations of the way seeds spread, Thoreau was stimulated by Darwin's book to deepen that investigation.

Jean Louis Rodolphe Agassiz, born in Switzerland in 1807, ten years before Thoreau, became a famous scientist at a young age. In his early career, Agassiz had studied fish and published books about extinct species of fish. Although he was aware of Darwin's work on the evolution of species, he did not accept that idea. He believed that species were special creations of God, and he therefore believed in the spontaneous generation of life. One piece of evidence brought up by those who believed in spontaneous generation was the way in which trees and other green plants appeared to spring from the ground where there did not appear to be seeds.

Thoreau was skeptical of this theory, and as was typical, he did not take someone else's word for it but attempted to find out for himself. He made direct observations about the ways in which seeds spread. He established that seeds were transported away from the parent plant by animals and wind, at times giving the superficial appearance that plants had arisen spontaneously. He saw seeds carried in the fur of small mammals, transported by birds, buried by squirrels. Instead of finding evidence to support spontaneous generation of life, Thoreau established that all trees in the forests near Walden Pond always germinated from seeds or resprouted from existing roots, never arising spontaneously from the soil.

In observing the way seeds were dispersed and germinated, Thoreau saw how the seeds of some trees were blown about by the wind. He looked at caches of seeds in animal burrows. He watched squirrels cut twigs of pitch pine heavy with cones and move them to new locations. "I counted this year twenty under one tree, and they were to be seen in all pitch-pine woods," he wrote of the squirrels. In contemplating this and other observations, he wrote, "Anything so universal and regular, wherever the larger squirrels and pitch pine are found, cannot be the result of accident or freak." He continued his observations and confirmed for himself that "squirrels were carrying off these pine boughs with their fruit to a more convenient place either to eat at once or store up." In the course of this transport, some of the seeds were abandoned and sprouted, spreading the species to new locations.

Thoreau did not stop with qualitative observations. He measured distances between trees of a given species. He counted and weighed seeds. He measured the distance over which nesting birds and squirrels carried fruits and seeds and conducted experiments to determine how far a bird could move a specific seed. He studied the dispersal of seeds that stuck to the fur of mammals, that were ingested by birds and mammals and released in their droppings, and that were transported by wind and water. Bradley P. Dean, editor of *Faith in a Seed,* Thoreau's posthumously published manuscript on the dispersal of seeds, suggested that Thoreau may have been "the first Anglo-American field ecologist to be influenced by Darwin's theory."

Thoreau was able to do this work within walking distance of his home in Concord. The depth of his knowledge of physical nature increased greatly from studies of details he observed locally; he did not have to travel to a vast, faraway wilderness for this. Knowledge of the processes of nature—part of the *tangible* characteristics of nature—was more easily obtained in the heavily altered countryside near Thoreau's home than in pristine wilderness.

Thoreau found the study of the dispersal of seeds not only intellectually challenging but also fun. "It is pleasant to observe any growth in a wood," he wrote. "There is a tract northeast of Beck Stow's Swamp, where some years since I used to go a-black-berrying and observed that the pitch pines were beginning to come in; and I have frequently noticed since how fairly they grew, clothing the plain as evenly as if dispersed by art." The regeneration was beautiful and symmetrical. "At first the young pines lined each side of the path like a palisade," he wrote.

He observed how forests became established and developed over time. He discovered that small seedlings and saplings of the pines also served to provide shade and protect germinating seeds, functioning as "nurse trees." He wrote a report for the state of Massachusetts on the succession of forest trees, one of the first uses of that term.

Thoreau's approach to the study of seed dispersal was consistent with his work as a surveyor and his studies of Walden Pond. In all these projects, he displayed innovativeness and an affinity for accurate observation and measurement, two important qualities in making contact with and knowing nature.

At the time of his death, in 1862, Thoreau had begun work on "The Dispersion of Seeds," the manuscript recently published as *Faith in a Seed.* He wrote that "the shade of a dense pine wood is more unfavorable to the springing up of pines of the same species than of oaks within it, though the former may come up abundantly when the pines are cut." He discussed how the early successional pines serve as "nurse trees" for the oaks. This was based on his observations as well as practices in England where oaks were planted under pines to benefit the oaks. Young oaks tend to grow well in the shade of pines, which protect them from the winter cold and the drying effects of direct sunlight. Thus, pines appear first in open fields, followed by oaks. Pines do not germinate in the deep shade of a dense forest. Species succeed one another.

Thoreau presented these observations and generalizations in a direct and objective manner, without expressing a strong preference for one kind of species over another regarding its role in nature or making moral judgments about the relative merits of pines and oaks.

In contrast, modern environmental issues attract a wide range of opinions, emotions, and moral judgments about species. Modern environmentalism is a sociopolitical movement with a wide range of perspectives and ideologies. Some environmentalists believe that solutions to environmental problems lie in gradual change. Others, frustrated by modern civilization, believe in radical change and a great diminishment of civilization, if not its complete elimination, to save nature. Still others believe quite the opposite, that environmentalism is no more than a political attempt to destroy civilization. What appears at first glance to be an argument over one of many political issues, how best to preserve the environment, turns out to touch on the deepest issues that confront human beings and society: the compatibility of civilization and nature and the deep meaning of nature to people.

The depth of the implications of these issues is illustrated by the ideas of the philosophical-political movement known as deep ecology, which proposes a radical solution based on fundamental change in the moral order. Deep ecology presents a critique of Western civilization fundamentally different from the ideas of most environmental organizations and activists, who seek to improve the environment within the context of Western civilization. In contrast, Arne Naess, one of the principal philosophers of deep ecology, succinctly prescribes a major change in the entire moral order, a change that fundamentally rejects the traditions of Western civilization. First, he describes the existing moral order: "Many contend that living beings can be ranked according to their *relative intrinsic value.*"

The classical ranking is that "if a being has an eternal soul, this being is of greater intrinsic value than one which has a time-limited or no soul." Then, "if a being can reason, it has greater value than one which does not have reason or is unreasonable." And "if a being is conscious of itself and of its possibilities to choose, it is of greater value than one which lacks such consciousness." Naess argues that "none of these standpoints" is justified. Instead, "they fade after reflection and confrontation with the basic intuitions of the unity of life and the right to live and blossom." He states that "the right of all the forms [of life] to live is a universal right which cannot be quantified. No single species of living being has more of this particular right to live and unfold than any other species."

Naess then suggests "a rather provoking thought experiment," which is that human beings recommend their own withdrawal as the dominant organisms on the Earth. In doing this, "the human drive for self-realization requires us to give way for the more perfect."

Naess states that "to relate all value [of other living things] to mankind is a form of anthropocentrism which is not philosophically tenable." He seems to argue that civilization, with its advanced technology, cannot coexist with nature. "The need for outdoor life and the need for machine-oriented technical unfolding cannot take place *simultaneously* [emphasis added]," he continues. Civilization harms nature, and people should willingly and consciously withdraw themselves, which means withdrawing also their creativity, art, poetry, music, philosophies, and philosophers.

Naess describes our species functioning ecologically as an early successional, pioneering species, with this term used in a derogatory sense. "Mankind during the last nine thousand years has conducted itself like a *pioneer invading species* [emphasis added]," he states. "These species are individualistic, aggressive, and hustling. They attempt to exterminate or suppress other species. They discover new ways to live under unfavorable external conditions—admirable!—but they are ultimately self-destructive. They are replaced by other species which are better suited to reestablish and mature the ecosystem."

Contradicting his previous argument that all living things are morally equal, Naess describes pioneering species disparagingly and by analogy condemns, or at best rejects, civilization, as well as our species, *Homo sapiens*. In so doing, Naess restates the great ancient, prescientific myth of the balance of nature, which was translated in the twentieth century as a rigid interpretation of ecological succession—the development of an ecosystem—as a process that leads from pioneering species to mature species, with the mature species forming ecosystems that persist indefinitely.

Naess argues that it is a moral good for *Homo sapiens* to transform itself from a pioneering species to a mature species—a highly anthropomorphic argument reflecting a significant misunderstanding of modern ecology. Today, ecologists understand, as did Thoreau, that different species are adapted to conditions created at different stages in the recovery of an ecosystem from a disturbance. If there is a stabilization of some kind in a forest or other ecological system, it is the result of systemic characteristics—of a system with both "pioneering" and "mature" species. Each species has a role in this unfolding drama, told and retold on the landscape. "Pioneering" species play a positive role, stabilizing the soil and taking up nutrients before they are lost to erosion after a disturbance, such as a fire in Yellowstone National Park or the eruption of Mount St. Helens.

In his study of the spread of seeds and the succession of forests, Thoreau did not reach a moral conclusion about the relative "sinfulness" or "purity" of either oaks or pines, nor did he moralize about pioneering or late successional species in general. He simply described the natural dynamics of forest recovery and conditions that favor or disfavor species adapted to different stages in this process.

On the surface, deep ecology appears to be a well-intentioned and well-articulated, if anti-human and anti-humanity, extension of prior rationales for protection of the environment. But its arguments are much deeper and more important than simply a debate about biological diversity, sustainable forests, or other specific topics related to biological nature. Sorbonne professor of philosophy Luc Ferry, in his recent book *The New Ecological Order*, argues that deep ecology, generally little known to the American public and frequently dismissed by environmental experts as a strange political-ideological fringe movement, presents the first significant opposition in some three hundred years to the philosophical tradition that has dominated Western civilization since the time of French mathematician and philosopher René Descartes. Ferry characterizes this tradition as emphasizing individuality, humanism, reason, rationality, and democracy—the very characteristics Thoreau appreciated. People are viewed as being at the top of nature's moral hierarchy because we can think, suffer, enjoy, create, and study. Nature is to be preserved because it helps us directly or indirectly, because its beauty is important to us, or because *we* believe it has a right to exist and therefore feel better knowing we are doing our best to help it exist. All these rationales are human centered.

In contrast, Ferry notes, the deep ecology movement begins with the premise that the persistence of life is a property of ecosystems and of the biosphere, the Earth's global life-supporting and life-containing system. This, too, is new knowledge and part of the new ecology. The global impact of life is a discovery of the past three decades of environmental sciences, but it has rapidly led to strong ethical, moral, and religious interpretations. As interpreted by the proponents of radical environmentalism, the conclusion is a surprise: what Ferry calls a "strange hierarchy." The simple notion that the whole (the biosphere) is more important than the parts (ecosystems, species, populations, individuals) leads to the deduction that "the totality is morally superior to individuals." The result of this apparently simple and innocent premise is a complete inversion of our moral order. The biosphere—an abstracted global system—is at the top, perceived as a "quasi-divine entity"; next are the nonsentient, nonrational forms of life that are innocent of intentional evil and are simply trying to survive, and at the bottom is *Homo sapiens,* doomed by its very rationality, which is seen as an original sin because it makes human beings uniquely capable of messing up the biosphere and therefore morally inferior to the likes of cockroaches, parasitic worms, and bacteria.

As Ferry points out, deep ecology is fundamentally different from the wide range of ideologies that are considered environmentalist but are philosophically within the traditions of Western civilization. This is why I emphasize the ideas of deep ecology—not because they represent the mainstream of environmentalism but because they represent a fundamental counter to the traditions of Western civilization. As such, they demand careful consideration. Deep ecology challenges fundamental assumptions of Western civilization, and for this reason it must be taken seriously in the debate about civilization and nature, precisely because it is fundamentally a movement against civilization. Deep ecology denounces the Judeo-Christian tradition and Platonic dualism, both of which place the human spirit and rationality above nature, and it also denounces science and our entire industrial-technological society. The legal and political implications of deep ecology are powerful. Naess, for example, argues that "the arrogance of stewardship [as found in the Bible] consists in the idea of superiority which underlies the thought that we exist to watch over nature like a highly respected middleman between the Creator and the Creation."

Believers, as reported by Ferry, dream of a "global government that can subjugate populations in order to reduce pollution and alter desires and behaviors through psychological manipulation." Proponents argue for trying people for crimes against nature. This, in turn, implies an anti-individualistic, anti-democratic political ideology wherein the individual, at the bottom of the moral pyramid, must be sacrificed for the good of the whole and our species must be sacrificed for the persistence of all other life on Earth and the Earth system that makes life possible. The future of the world, the deep ecologists say, lies in the suppression of individualism and democracy so that the biosphere can persist. Not only are people who commit crimes against nature evil, but also massive die-offs of people are good. Ferry quotes one of the radical ecologists, William Aiken, who asks whether it is, perhaps, our duty to create these die-offs. People and civilization are to be sacrificed because it is we and our creativity that are destroying the biosphere. Civilization kills nature, and therefore it must be stopped.

Deep ecology also leads to an argument for preserving large areas of wilderness, defined in the modern sense as places free of human influence. "The deep ecology demand for the establishment of large territories free from human development has recently gained in acceptance," notes Naess. "It is now clear that the hundreds of millions of years of evolution of mammals and especially of large, territory-demanding animals will come to a halt if large areas of wilderness are not established and protected."

A strange irony arises: deep ecology rejects as evil the very rationality and individualism that allowed science to grow and flourish, yet it is that science which led to the understanding that life is sustained by ecological systems, that life has affected the Earth's environment at a global level for several billion years, and that life depends for its persistence on the global life-support system. The same science that made possible the premise of the new radical environmentalism is rejected as evil.

An equal threat to achieving the vision of this book—a flourishing of both humanity and nature, and a compatibility between civilization and nature—is the perspective of those in the "wise use" movement and related vocal minorities who do not believe that civilization needs to limit itself to any extent because of nature, that human action can continue unfettered on the globe. Because they believe this, they perceive the actions of environmentalists as no more than a subversion of civilization. Recently, I attended a conference held in International Falls, Minnesota, a town surrounded by wetlands, lakes, streams, and cutover and regrown forests—a westward extension of the same boreal forest that grows in the Maine woods. The president of the Society of American Foresters at that time said in a speech that environmentalism was anti-Christian and anti-capitalist. It was shocking to hear the head of a major organization, whose journal publishes scientific articles about forestry, make such a statement. Although such opinions are in the minority, they are strongly held.

Ron Arnold's book *Ecology Wars* has been called "the bible of the wise use movement." However, the book is a discussion of environmentalism as an activist social and political movement and deals little with ideas about the functioning of biological nature. "Environmentalism is an institutionalized movement of certain people with a certain ideology about man and nature," Arnold asserts, and "the goal of our ecology wars should be to defeat environmentalism." He separates the environment from environmentalism and states that the latter "is the excess baggage of anti technology, of anti-civilization, of anti-humanity, of institutionalized lust for political power that we must reject."

It would have been helpful if Arnold had been explicit in his book about the scientific information on which he based his conclusions, to permit a direct comparison between the wise use movement and deep ecology. But he gives little attention to scientific information or the condition of the environment and nature, limiting his discussion to statements such as that pollution "is an inevitable consequence of life at work. In a sensible world industrial waste would not be banned but put to good use." He does briefly discuss the Earth's life-supporting system. First, he quotes James Lovelock, originator of the Gaia hypothesis—the idea that life affects the Earth's environment at a global level and depends on that global system. Lovelock repeats what G. Evelyn Hutchinson, one of the major ecologists of the twentieth century, had written before: the interesting thing about the Earth's surface is that the atmosphere, oceans, and continents are not in a chemical equilibrium. By this, Lovelock and Hutchinson mean that if the Earth's atmosphere, oceans, and continents were isolated in a closed box and left alone without any input of sunlight or other energy, or even if they were open to the flow of energy that is simply reflected or absorbed by the surface as a physicochemical system, the composition of the three would come into a steady state. That steady state would include an atmosphere much like those of Venus and Mars and unlike that of Earth, with little or no free oxygen or free nitrogen, composed mainly of carbon dioxide and nitrogen compounds such as ammonia. Life, Hutchinson pointed out, pushes this system away from a chemical equilibrium. Lovelock added that the disequilibrium not only is the result of biological activity but also seems "more like a biological construction: not living, but like a cat's fur, a bird's feathers, or the paper of a wasp's nest, an extension of a living system designed to maintain a chosen environment."

Even more remarkable to Lovelock, the conditions appear to be maintained at "optimum values from which even small departures could have disastrous consequences for life." Lovelock concluded that life has affected the Earth for billions of years and that these effects have tended to maintain environmental conditions in optimal states for life.

Arnold quotes a passage from Lovelock that makes some of these statements, and then he comments that "Lovelock's clear-sighted vision of a self-protecting Earth managed for ages by self-knowing human stewards slaps the doctrinaire environmentalist vanguard squarely in the political philosophy." Here, Arnold confuses Lovelock's ideas about life's very long term effects on the planetary environment with his ideas about human stewardship.

Arnold's book is a handbook about how to defeat environmentalism, if one accepts environmentalism as bad. However, except in a few passages, he does not address scientific information directly. From Arnold's extreme viewpoint, environmentalism is an evil sociopolitical force that must be opposed and defeated in order to save civilization and the free enterprise system. He states briefly that "we can love the Earth and its community of life without hating technology, without wanting to destroy industrial civilization, without wallowing in an orgy of self-loathing."

It is interesting that both books—Arnold's on the wise use movement and Naess's on deep ecology—turn to new scientific discoveries about life's effects on our planet's environment in defense of their arguments. Both authors misunderstand scientific information and then arrive at conclusions based on their misunderstanding, which are in turn used as justification for their ideologies. Both begin with an ideology and are political and social in focus. Each sees the "other side" as evil and dangerous to individuals and to the world. Both sides blame governments.

Both see the battle as about who is on the side of the right ideas and who is on the side of the wrong ones. Their ideas are fixed, so investigation and consideration of facts, data, and information are irrelevant.

These points of view articulate the problem: Must we choose between civilization and nature, or can we have both? Are they compatible? Most Americans are in the middle and state that they like a good environment. Over the past twenty years, Americans have consistently ranked environmental concerns among the top ten issues. In public opinion polls taken in 1998, two-thirds of respondents considered themselves strong to moderate environmentalists and more than half of the population believed that the state of the environment had worsened in the past two to three years.

The public is also aware of the holes in the arguments posed by deep ecologists and proponents of the wise use movement. Polls show that the majority of people believe that public discussions of environmental issues are "not honest" for two reasons: adequate information is lacking, and existing information is presented in a biased manner, to favor specific political and ideological viewpoints. That is, we do not understand nature as well as we need to, and when we do understand it, there are people ready to misinterpret the information to serve a political or ideological goal.

The question, then, is whether it is possible to articulate another approach, a different way of thinking. The issue for a rational, well-meaning, and prudent person is how to increase the probability of the combined persistence of nature and civilization and to maintain for human beings both physical and spiritual well-being. Part of the solution is to observe nature carefully and neutrally, as Thoreau did in making his careful, detailed observations of the spread of seeds and the succession of forest trees.

Another part of the solution is to see nature and civilization as one system, not two. This is why Thoreau's lifelong fascination with swamps is an important metaphor. His perception of wilderness and the relationship between human beings and nature goes to the heart of the major issues confronting civilization today: whether humanity's effects on the environment are so great as to threaten the Earth's entire life-supporting system and whether we are morally obligated to throw aside individualism, rationalism, democracy, and civilization as we know it in order to save the environment.

Rather than seeing species or ecosystems as important in themselves, Thoreau saw nature as important to human creativity, civilization, and culture. A civilization must keep its swamps if it is to survive, but it cannot focus only on swamps. We can save nature from the problems that confront it only if we maintain the very civilization that has given us the tools to understand and deal with these problems. Goodwill toward nature is a product of the human psyche. It will not blossom if civilization declines and human populations remain large: an uncivilized, overcrowded populace would battle over nature's resources without the buffering effects of a social order.

The swamp on the edge of town provides the mixture that leads to the best in human life, and that was Thoreau's desire. Nature and civilization are integrated; they need each other. Perhaps if we learn to "think like a swamp," we might find a way to sustain both the organic soils and the organic innovation of humanity.

Chapter 5

Racing in the Wilderness

*"Thoreau did not think of the wild—or of walking either—
as a special preserve. It was not for recreation so much as it was
for re-creation."*

Robert Sattelmeyer, *The Natural History Essays*

*"Barring love and war, few enterprises are undertaken with such abandon,
or by such diverse individuals, or with so paradoxical a mixture of appetite
and altruism, as that group of avocations known as outdoor recreation."*

Aldo Leopold, *A Sand County Almanac*

ON JULY 31, 1857, four days after he was lost in Umbazooksus Swamp, Henry David Thoreau was nearing the end of his third and last trip to the Maine woods. He, his companion, Ed Hoar, and his Indian guide, Joe Polis, had canoed on the Allagash River and the East Branch of the Penobscot River. The going had been difficult down the East Branch of the Penobscot. Thoreau had "spent at least half the time in walking" and wrote that "the walking was as bad as usual," probably because of dense underbrush along the shore. The party stopped to camp "about a mile above Hunt's, which is on the east bank, and is the last house for those who ascend Ktaadn on this side."

On the next day, August 1, they went down the river in the canoe and "stopped early and dined on the east side of a small expansion of the river, just above what are probably called Whetstone Falls, about a dozen miles below Hunt's." The Hunt House, an inn built in the 1830s, was closed because the owners were away. As a result, Thoreau and his companions were unable to obtain supplies there. Being near the end of their trip, they were probably tired and low on supplies. It was a time in a trip when it would be easy to become testy and unhappy.

From this camp, they portaged about three-fourths of a mile around the falls, where the "rocks were on their edges, and very sharp," Thoreau wrote. "When we had carried over one load, the Indian returned by the shore, and I by the path, and though I made no particular haste, I was nevertheless surprised to find him at the other end as soon as I. It was remarkable how easily he got along over the worst ground."

As a result, Thoreau and Polis decided on a race. "He said to me," wrote Thoreau, "'I take canoe and you take the rest, suppose you can keep along with me?'" Thoreau explained: "I thought that he meant, that while he ran down the rapids I should keep along the shore and be ready to assist him from time to time, as I had done before; but as the walking would be very bad, I answered, 'I suppose you will go too fast for me, but I will try.'

"But I was to go by the path, he said. This I thought would not help the matter, I should have so far to go to get to the river-side when he wanted me. But neither was this what he meant. He was proposing a race over the carry, and asked me if I thought I could keep along with him by the same path, adding that I must be pretty smart to do it. As his load, the canoe would be much the heaviest and bulkiest, though the simplest, I thought that I would be able to do it, and said that I would try. So I proceeded to gather up the gun, axe, paddle, kettle, frying-pan, plates, dippers, carpets, etc., etc., and while I was thus engaged he threw me his cowhide boots.

"'What, are these in the bargain?' I asked. 'Oh, yer,' said he; but before I could make a bundle of my load I saw him disappearing over a hill with the canoe on his head, so, hastily scraping the various articles together, I started on the run, and immediately went by him in the bushes, but I had no sooner left him out of sight in a rocky hollow, than the greasy plates, dippers, etc., took to themselves wings, and while I was employed in gathering them up again, he went by me; but hastily pressing the sooty kettle to my side, I started once more, and soon passing him again, I saw him no more on the carry. I do not mention this as anything of a feat, for it was but poor running on my part, and he was

obliged to move with great caution for fear of breaking his canoe as well as his neck. When he made his appearance, puffing and panting like myself, in answer to my inquiries where he had been, he said, 'Locks (rocks) cut 'em feet,' and laughing added, 'Oh, me love to play sometimes.'"

Polis told Thoreau that he and his companions often held such informal races. Meanwhile, as a result of the race, Thoreau noted, "I bore the sign of the kettle on my brown linen sack for the rest of the voyage."

When he was not lost in a swamp or ascending Mount Katahdin, Thoreau had fun traveling over large distances in the big woods. J. Parker Huber, who retraced all of Thoreau's Maine travels, walking where Thoreau walked and canoeing where he canoed, and who read carefully everything Thoreau wrote in his journals about his Maine travels, commented: "It is evident that he had a lot of fun in Maine. He laughed at himself and at others. Banter between Polis and Thoreau continued night and day. Their peals of laughter resounding across the lakes must have cheered even the loons."

Laughter and sheer enjoyment or pleasure are very human qualities that are perhaps one of the roots of creativity, and they certainly are roots of curiosity, given that we often become curious about the things we enjoy. Had Thoreau not had fun—enjoyed himself—in the woods, would he have studied the woods as much as he did? Would he have sought to understand his connection to nature both spiritually and physically? Would he have found it a source of creativity, the swamp a necessary place for poets to visit? These questions are worth contemplating as we confront the heavy-handedness of our technologies and our moralizing about nature.

Thoreau's journal entries about his walks in the woods around Concord reflect the same sense of fun. In the winter, he enjoyed skating and sliding on the river's ice. His notes about what he did on Christmas Day in various years are good illustrations of his sense of fun in the outdoors. On December 25, 1853, when he was thirty-six, Thoreau went for a walk near

Concord and wrote: "Staked to Fair Haven and above. . . . About 4 P.M. the sun sunk behind a cloud, and the pond began to boom or whoop. . . . It is a sort of belching, and, . . . somewhat frog-like. . . . It is a very pleasing phenomenon, so dependent on the altitude of the sun." A few years later, on December 25, 1857, when he was forty, he wrote that he had "skate[d] on Goose Pond," and on Christmas Day the next year, he wrote: "The ice on the river is about half covered with light snow. . . . I go running and sliding from one such snow-patch to another. . . . It is so rough that it is but poor sliding withal." Here was a forty-one-year-old man spending Christmas Day slipping and sliding among the snow patches on the ice-covered river. For Thoreau, time spent in the woods near Concord was recreation, fun, joy; traveling in the Maine woods, the big woods, was adventurous, exciting, and fun. This is not the image of Thoreau that most people have formed on the basis of hearing a few quotations about the importance of nature or from reading *Walden.*

Thoreau's search on Mount Katahdin makes clear that he valued nature and sought spiritual contact with it. Although he did not find the hoped-for contact on the mountain, he did elsewhere. Often, this kind of contact with nature occurred for Thoreau as a result of careful examination of the small and the detailed in nature, both in the Maine woods and in his walks around Concord. During his last trip to the Maine woods, on the night of Friday, July 24, 1857—before he was lost in the swamp—while camping in "a dense and damp spruce and fir wood," Thoreau woke during the night and saw pieces of wood glowing in the dark—phosphorescent wood—which he had heard about but never seen. The glow, which he found on small twigs and part of a decayed stump, was "fully as bright as the fire" but white rather than yellow or red. He attributed it to "the previous day's rain and long-continued wet weather."

Thoreau examined the twigs closely, as an enthusiastic naturalist would, and found "that the light proceeded from that portion of the sap-wood immediately under the bark, and thus presented a regular ring at the end." He stripped back the bark, cutting "into the sap," and found "it was all aglow along the log" although the log itself was still solid. "It could hardly have thrilled me more if it had taken the form of letters, or of the human face," he wrote. "It made a believer of me more than before. I believed that the woods were not tenantless, but choked full of honest spirits as good as myself any day,—not an empty chamber, in which chemistry was left to work alone, but an inhabited house—and for a few moments I enjoyed fellowship with them." This small phenomenon raised in Thoreau a sense of wonder and contact with nature, a spiritual feeling. He made clear that this experience was important, but not in terms of tangible, physical knowledge of nature. "I let science slide," he wrote, "and rejoiced in that light as if it had been a fellow-creature. . . . A scientific explanation, as it is called, would have been altogether out of place there. That is for pale daylight. Science with its retorts would have put me to sleep; it was the opportunity to be ignorant that I improved."

Another intangible quality that Thoreau found in nature, especially in his image of the swamp, was nature as inspirational to human creativity, a necessity for poets, so to speak.

These intangible qualities—joy, spirituality, and creativity—seem difficult reasons to put forward publicly for valuing nature and acting to conserve it. In public justifications for conserving nature, there is a tendency to focus on material benefits: utilitarian benefits or products of direct commercial value that can be obtained from species living in natural ecosystems, and public service benefits of natural ecosystems. An example of a utilitarian benefit is the rubber from the tropical rain-forest tree that was used to produce Thoreau's bags. Public service benefits include the fixation of atmospheric nitrogen by bacteria in soils and open waters; the fertilization of flowers by bees, birds, and bats; the release of oxygen into the atmosphere and removal of carbon dioxide by photosynthetic organisms; and the retardation of soil erosion by trees and other vegetation.

Several years ago, I attended a lecture about conservation of life on Earth by an expert on biological diversity. He gave four reasons for conserving nature—utilitarian, ecological, aesthetic, and moral—but then spent the rest of his time discussing the utilitarian benefits, specifically the potential for finding new pharmaceuticals in the New World Tropics and his work with Indians of the Amazon Basin to help them obtain a fair share of royalties from any such discoveries. The emphasis was thus on direct economic benefits. Similarly, a film I saw recently at the visitors' center of Yellowstone National Park showed the beauty of the park and its impressive wildlife but gave special attention to an enzyme derived from bacteria that live in the park's hot springs, emphasizing that the enzyme is fundamental to all genetic engineering.

Such material benefits are certainly reasonable justifications for conserving nature, but standing alone they can lead to curious conflicts. This became evident when the drug Taxol was discovered in the Pacific yew, a small tree that grows in forests of the Pacific Northwest. Taxol appeared to be beneficial in treating certain cancers, providing a utilitarian reason to conserve the natural ecosystem of the Pacific yew. But getting Taxol seemed to require cutting or killing Pacific yew trees. As soon as this became clear, some environmentalists reacted negatively, arguing that the Pacific yew should not be sacrificed or threatened with extinction even though it might contain a product that cured cancer. This conflict was resolved happily when laboratory scientists developed a synthetic version of Taxol—but, of course, that development eliminated the utilitarian benefits of wild Pacific yew trees, which therefore served only as a source of discovery, not as a means of supply.

The dispute over Taxol revealed apparently contradictory rationales. On the one hand, the public had been told that the potential for discovery of such products was exactly the reason to conserve nature. On the other hand, when such an opportunity arose, the public was informed that people should keep their hands off the Pacific yew because its harvest might threaten the persistence of that species. This seemed to reveal an underlying desire to conserve the Pacific yew for reasons that had to do not with commercial or public service benefits but with something else. That "something else" is rooted in intangible values such as joy, creativity, and spirituality—values that Thoreau understood.

Modern environmental textbooks and other environmental literature often list aesthetics as a reason to conserve nature, but this usually refers to scenic beauty—the physical appearance of scenery as captured in photographs, with little, if anything, said of its spiritual or inspirational value. Today, when a moral justification is given for the preservation of nature—of the kind discussed by philosopher Arne Naess and many others—it is that nonhuman life has an intrinsic right to exist. This is more a legalistic than a religious or spiritual justification, although the motivation for it may be religious or spiritual. In contrast, Thoreau valued nature for its spiritual, creative, and religious qualities—he did not shy away from these intangibles.

Why do people who love nature tend to resort to material justifications for conserving it? I believe that it is because in our heavily materialistic society, there is a sense that nonmaterialistic arguments will not "sell," will fail to hold up against economic arguments. Many environmentalists have concluded that they must answer economic arguments against conserving nature with economic arguments on behalf of it. But in the retreat to materialistic justifications, there is an implicit retreat *from* a belief in the power of ideas and the power of human passions.

The two aspects of the duality of our experience, the physical and the spiritual, are intertwined in our connections with nature; each affects the other. But this connection is rarely discussed openly. Two people fishing for landlocked salmon on the Allagash River, for example, might do it for quite different motivations. One person might be there because fishing is recreational in the sense that golf and tennis are recreational: it is pleasant to be outdoors, but the primary motivations are the challenge of outwitting one's opponent—in this case, the fish—and the benefit of engaging in an activity that takes one's mind away from other cares. The other person might be there because fishing is a focused activity that brings one into contact with nature and provides an opportunity to better oneself. The detailed knowledge that the second person gains from fishing affects the intangible. Neither angler's motive is to be dismissed, but it is important to recognize their differences. Thoreau was like the second kind of fisherman.

Values can be difficult to discuss. Modern scientists generally avoid such discussions, and rightly so, for the human connection with values has to do with the metaphysical as well as the physical, and the metaphysical is outside the realm of science. Yet even scientists outside their professional roles—speaking or writing as individual human beings, as citizens—shy away from the subject. It is a dangerous one to write about because such a discussion can easily be taken the wrong way, interpreted as a scientist's attempt to integrate metaphysical statements into his professional work or to appear to be an expert on topics far outside his true area of expertise. More simply said, talking about values can make a scientist look like a flake.

But ideas have great power. Robert H. Nelson, a political scientist who specializes in natural resources, made the point succinctly: "Ideas are more important than many practical men and women of affairs believe. Ideas shape institutions and give them social legitimacy. The failure of an idea can in the long run mean the demise of an institution. Ideas motivate and define the culture of organizations. Without a clear idea of mission and purpose, an organization risks declining morale and employee commitment." If this is so, it is not necessary for environmentalists to hide their intangible reasons for appreciating nature behind materialistic justifications.

Moreover, the materialistic justifications are often faulty because science has so primitive an understanding of the workings of complex ecological systems. The result is confusion about why and how we should conserve nature. The public recognizes this. As mentioned earlier, public opinion polls indicate that most people believe that debates about the environment are not "honest," meaning that they are not based on legitimate facts or a strong understanding of nature.

There is one more rationale for conserving nature: the need to conserve human cultural heritage that is found only within a certain ecosystem or set of ecosystems. Again, Thoreau provides an example. During his third trip to the Maine woods, he visited Mount Kineo, an unusual isolated, rocky hill on the shores of Moosehead Lake. In his book *The Maine Woods,* Thoreau notes that the name of the lake supposedly comes from the shape of this mountain, which looks like a moose's head. While canoeing to the base of the mountain, Thoreau, Ed Hoar, and Joe Polis had difficulty finding a suitable landing and camping place. They settled on a camp that required them to hike "half a dozen rods" through dense fir and spruce "almost as dark as a cellar," he wrote. "It had been raining more or less for four or five days," but Polis found dry bark "from the underside of a dead leaning hemlock, which he said he could always do"—another piece of natural history Thoreau learned from Polis, useful in camping. The mountain extended 700 feet above the lake, Thoreau estimated, and was formed of an unusual mineral, "generally slate-colored, with white specks." Known at that time as hornstone and used in the past by Indians to make tools, it is now called rhyolite and is known to be a combination of quartz and orthoclase. It was formed in the Devonian period, approximately 375 million years ago.

In *The Maine Woods,* Thoreau cites a geological report stating that Mount Kineo is the largest mass of this material in the world. Thoreau was as interested in Indian artifacts as he was in Indian culture. "I have myself found hundreds of arrow-heads made of the same material," he wrote, having found some of these along the Concord River, where the Penobscot Indians had come to trade them long before Thoreau's time. According to J. Parker Huber, whose book *The Wildest Country* retraces Thoreau's Maine travels, "Indians came from all over Maine, if not from greater distances, for this ancient rhyolite. For them, Kineo was one of the richest sources of raw material in New England." Parker also noted that natural erosion "performed much of the work of shaping the stone for the Indians." Weathering eroded the mountain's southern face, causing the rock to break off and fall, splitting into many pieces of a size quite useful for toolmaking. While on Mount Kineo, Thoreau found a "small thin piece which had so sharp an edge that I used it as a dull knife." To try it out, he used it to cut an aspen twig an inch thick, inadvertently cutting his fingers as well.

Some people suggest that nature should be conserved in order to preserve the ecological and geological conditions that have allowed cultures with such technologies and practices to persist. This argument is put forward as one reason to conserve Amazonian rain forests. Some Sioux Indians also present the argument as a reason to establish migratory corridors for bison where not only the bison could roam but also Indians could practice their traditional cultural migration and hunting. This rationale came to the fore in 1999 when Pacific Northwest Indian tribes hunted and killed a gray whale, using a traditional canoe. They said that such practices were essential for the continuation of their culture.

Years before his race in the wilderness, when he and his brother built a boat and took a weeklong trip on the Concord and Merrimack Rivers, Thoreau wrote comments about people and nature that seem a little strange from a modern point of view. Describing one of the first nights of the trip, Thoreau wrote: "For the most part, there was no recognition of human life in the night, no human breathing was heard, only the breathing of the wind. At intervals we were serenaded by the song of a dreaming sparrow or the throttled cry of an owl, but after each sound there was a sudden pause, and deeper and more conscious silence, as if the intruder were aware that no life was rightfully abroad at that hour." But the passage continues: "There was a fire in Lowell [Massachusetts], we judged, this night, and we saw the horizon blazing, and heard the distant alarm bells. But the most constant and memorable sound of a summer's night was the barking of the house dogs more impressive than any music." This was "evidence of nature's health." From a modern perspective, it is curious that on a trip that seemed to be about the discovery of nature, the sounds of a town and the barking of dogs were to Thoreau "evidence of nature's health." Civilization and nature were intertwined; the heritage of civilization was tied together with the heritage of nature. Knowing one was part of knowing the other.

In this chapter, I have suggested eight rationales for conserving nature: recreational, spiritual, inspirational, utilitarian, ecological, aesthetic, moral, and cultural. Thoreau's rationales were human oriented. I found little if any discussion in his writings of an intrinsic value of nature independent of the ability of human beings to benefit from it. Thus, of the eight reasons to conserve nature, Thoreau would seem to have supported all but what is today called the moral.

Although Thoreau did not want to live permanently in the big woods of Maine and did not receive on the summit of Mount Katahdin a feeling that nature was benign and valued or cared about him, he thoroughly enjoyed himself in the woods. Such pleasure, one of the intangible values of nature, is a part of the development of creativity and inspiration from nature and a part of spiritual contact with it. Having fun helps us expand and clarify the reasons for conserving nature.

When we decide that we want to conserve nature, we have to be clear about our reasons and, through these, our goal or goals. Each goal requires its own path to understanding nature and the role of human activities in relation to it. If we do not choose a goal clearly and specifically, we will founder in our attempts to conserve nature and find ourselves caught in contradictions such as those regarding the Pacific yew. Moreover, we must be willing to state our real goals without fear that they will not have the political power of other rationales. We must not hide our real desires behind ones that we imagine will sell.

Chapter 6

On Horseback Confronting the Great Desert

"One day I rode to a large salt-lake. . . . A field of snow white . . . one of these brilliantly white and level expanses in the midst of the brown and desolate plain, offers an extraordinary spectacle."
Charles Darwin, *The Voyage of the Beagle*

"Let us faithfully record the impressions of the day, and depend upon it, both we and the world shall be wiser for it."
George Perkins Marsh, *The Camel*

ON AUGUST 3, 1846, a small party of men camped near a "sluggish spring." During the night, one of them wrote, they saw below them a great valley, lit "by the red glare of the moon, and the more pallid effulgence of the stars, to display imperfectly its broken and frightful barrenness, and its solemn desolation." They saw no living thing except themselves, "nor voice of animal, no hum of insect." In the countryside, "all was silence and death." Even the winds "seemed stagnant and paralyzed by the universal death around." They were without a guide "or any reliable index to [their] destination" but were about to travel seventy-five miles through a landscape without water or grass.

Lighting a fire in the gathering dawn, they packed their mules "with unusual care," for they were in a real wilderness, the salt desert near the Great Salt Lake in what is now the state of Utah. Stopping to repack their mules might cost so much time as to force them to camp in the desert an additional night, prolonging the time in which man and mule had no water to drink. They emptied a powder keg and filled it with water from a brackish spring, the best they could find. Then they rode six miles to a mountain summit, where, "shivering with cold," they started down the western slope through "straggling, stunted, and tempest-bowed cedars," eventually reaching a valley about ten miles wide and then a "ridge of low volcanic hills, thickly strewn with sharp fragments of basalts and a vitreous gravel resembling junk-bottle glass." From this summit they saw, stretching to the horizon, a desert of "snowy whiteness" like "a scene of wintry frosts and icy desolation" without a shrub or any sign of life—a "perfect hiatus" of life. They were looking at dried crystals from the Great Salt Lake, which evaporating waters had been depositing slowly since the end of the last ice age.

As they crossed the salty desert, their mules sometimes sank to their knees in the strange deposits, which seemed like a frozen river in a land of great heat. "About eleven o'clock," their chronicler wrote, "we struck a vast white plain, uniformly level, and utterly destitute of vegetation or any sign that shrub or plant had ever existed above its now-lake surface." The men rested the mules, moistened their mouths with brackish water, and began to ride again, entering into "a scene so entirely new to us, so frightfully forbidding and unearthly in its aspects, that all of us, I believe, though impressed with its sublimity, felt a slight shudder of apprehension."

Stranger still, as they rode, they began to see fifteen or twenty "figures of a number of men and horses," mounted and dismounted, far away but appearing to be gigantic. "Very soon, the fifteen or twenty figures were multiplied into three or four hundred, and appeared to be marching forward with the greatest action and speed." At first, they thought the men might be hostile Utah Indians, then perhaps a party of the army of Captain Fremont of California, but then, the author of this account "noticed a single figure apparently in front in advance of all the others" and was "struck with its likeness to myself." Realizing that the entire army rushing toward them might be a mirage, he stretched his arms out full length, turned his face to the side, and saw that this figure did exactly the same. He was facing himself; it was his own image that was hurtling toward him. Recognizing the figures for what they were, he and his companions still found that "this phantom population, springing out of the ground as it were, and arraying itself before us as we traversed this dreary and heaven-condemned waste . . . excited those superstitious emotions so natural to all mankind."

This truly dangerous adventure through American wilderness was experienced and chronicled by Edwin Bryant, a Massachusetts contemporary of Henry David Thoreau, in the same year Thoreau climbed Mount Katahdin. Bryant was born in 1805, twelve years before Thoreau, in Pelham, Massachusetts, a small village near Amherst, the present location of Amherst College, Hampshire College, and the University of Massachusetts. Early in his life, Bryant, like Thoreau, wished to write and to travel. In 1830, as a young man, he went to Louisville, Kentucky, where he went to work for the city's newspaper, the *Louisville Journal*. In 1834, he became editor of the *Lexington (Kentucky) Intelligencer*, and ten years later, he established his own paper, the *Louisville Morning Courier*.

In Louisville, he saw and heard about people traveling west and was caught up in the excitement. In 1846, he began a journey to California over a poorly defined trail. Like Thoreau, he was on a "literary mission," traveling with a party of men but planning to write a book that would give an accurate description of the West, including its geology and botany. After traveling by steamboat to Independence, Missouri, he joined a group of emigrants led by William H. Russell, with the goal of reaching California's Sacramento Valley.

Although he is little known today, Bryant was well known in his own time. The publication of his book, *What I Saw in California*, in 1848 was perfectly timed: it became the principal guidebook for those traveling to California to join the gold rush. Unlike Thoreau's early works, which sold poorly in their time, *What I Saw* was an instant best seller.

Bryant and Thoreau had similar goals: experiencing and writing about America's nature. Both men were struck by the physical nature they observed; both were influenced by the spiritual tie they felt with nature. They were active at the same time. While Bryant was crossing the American West, in late July 1846, Thoreau was writing at Walden and had been living there for a year. On July 23 and 24 of that year, Thoreau was detained in Concord for nonpayment of the poll tax. He spent the night of July 23 in jail, one of the most famous experiences of his life. Thoreau's act of civil disobedience reflected his concern with human society and civilization, which were intimately connected to his concern with nature.

Bryant, meanwhile, was traveling through rugged mountains in Utah near the course of Weber River, which flows into the Great Salt Lake. On July 24, he and his companions and their baggage-carrying mules followed what he called a small Indian trail that wound over and under cliffs. Reaching a summit, they walked for two miles "in a path so narrow that a slight jostle would have cast us over a precipice to the bottom of a gulf a thousand feet in depth." They followed a stream for another five miles and reached a grassy valley with a meandering stream and then the banks of Weber River, where they camped. The next day, July 25, Bryant noted that the valley was about fifteen miles long and one to three miles wide, with "mountains on both sides" rising, "in benches one above another," several thousand feet to snow-covered slopes. "It was scarcely possible to imagine a landscape blending more variety, beauty, and sublimity," he wrote, especially because of "the quiet, secluded valley, with its luxuriant grass waving in the breeze, the gentle streamlet winding through it as well as the wild currant with its ripe fruit, the trembling aspen, and the striking topography."

Earlier in his journey, before Bryant confronted the Great Salt Desert, he had passed through the prairies along the Kansas River, where one morning he saw "a solitary wild rose, the first I have seen blooming in the prairies," whose "delightful fragrance . . . excited emotions of sadness and tenderness, by reviving in the memory a thousand associations connected with home, and friends, and civilization, all of which we had left behind, for a weary journey through a desolate wilderness." He wrote: "It is not possible to describe the effect upon the sensibilities produced by this modest and lonely flower. The perfume exhaled from its petals and enriching the 'desert air,' addressed a language to the heart more thrilling than the plaintive and impassioned accents from the inspired voice of music or poesy." Like Thoreau, Bryant appreciated civilization as well as the beauty of the wild American landscape.

Bryant appreciated much the same things in nature and civilization as did Thoreau, but he experienced a much greater range of wilderness, from desert to grasslands, from mountain forests to the coast of California. Camping in the Black Hills of what is now South Dakota, Bryant appreciated both their beauty and their geological history.

On July 1, 1846, he camped in a "small oval-shaped valley," where he appreciated "a rivulet of pure, limpid water." He observed a thick scattering of volcanic debris and noted, "Many ages ago, the spot where we

are encamped, and where the grass is now growing, was the creator of a volcano; but its torch is extinguished forever." Bryant had a strong imagination and was aware of the new discoveries about geological history, including fossils of animals. He wrote that "where then flowed the river of liquid fire, carbonizing and vitrifying the surrounding districts, now gurgles the cool, limpid current of the brook . . . the thunders of its convulsions, breaking the granite crust of the globe, upheaving and overturning the mountains . . . are now silenced." The explosions that had once "affrighted" the "huge monster animals which then existed" were now dissipated.

It was at the end of this summer that Thoreau left Walden for one of his trips to the Maine woods, where, in August 1846, he reached North Twin Lake and saw by moonlight a landscape "completely surrounded by the forest as savage and impassible now as to the first adventurers." The landscape, he wrote, had "a smack of wildness about it as I had never tasted before." While Bryant was struggling through a true wilderness, much of it inhabited sparsely by Indians but with mile after mile devoid of other people, Thoreau was writing that "it is difficult to conceive of a country uninhabited by men—we naturally suppose them on the horizon everywhere—and yet we have not seen nature unless we have once seen her thus vast and grim and drear—whether in the wilderness or in the midst of cities."

It is curious, therefore, that Bryant's best-seller of 1848 is known no longer, whereas Thoreau's writings, generally poorly selling works during his lifetime, are so influential today. Why is this so? It is a difficult call. The simple answer appears to be the overriding importance of our spiritual quest for a connection to nature. Although Bryant's physical experiences in wilderness across the American West impressed him deeply, affecting his sense of connection with nature, in his book he does not dwell on the broader and deeper philosophical and religious implications of what he saw and felt. These two Massachusetts natives shared much in common: a concern with writing about nature, a love of both natural scenery and civilization, and an inner response to the beauties and sublimities of the American wilderness. By most standard methods of evaluation—clarity of style, coherence and consistency in the flow of material, success in sales, and reputation of the book among his contemporaries—Bryant's first book was better and more successful than Thoreau's first work, *A Week on the Concord and Merrimack Rivers.* It does not suffer the clumsy expressions or sophomoric pretensions at philosophy of Thoreau's first book. But Thoreau's later works have completely eclipsed Bryant's one book.

Ironically, we today, who believe we live in a practical age—when many write that the solutions to environmental issues and our concern with nature are only a matter of technical details—are more impressed with the writings of Thoreau, who wrote more about the meaning of what he saw than about the facts of the scenery. Like Meriwether Lewis and William Clark, Bryant was at his best describing what the countryside was like and how he moved from one place to another. He told his readers clearly and with passion of both the beauty and the hardships of the trail to the California coast. His is the stuff of adventure and wilderness experience. Thoreau, meanwhile, traveling with good Indian guides through less wild, often cutover, and heavily used landscapes, spoke to us about what it was like to be there and was able, through his imagination, to relate to us the inner meaning of his outward experiences. This difference is all the more striking because Thoreau was not only a good naturalist but also a great defender of civilization and human culture, unlike many late-twentieth-century natural history writers who wrote only of the destructive powers of human beings and technology. The contrast between the lasting effects of Bryant's and Thoreau's writings seems consistent with the idea that a broad combination of motivations to conserve nature—spiritual as well as physical—is what moves people today, as it always has done. A practical guidebook may be more popular in the short run, but a contemplation of human beings' deeper, ancient concerns with the meaning of their existence, so dependent on the connections between humanity and nature, has greater staying power.

Chapter 7

Measuring the Pond

"It ain't what you don't know that gets you,
it's what you do know that ain't true."
Attributed to Will Rogers

"Much has been learned since the end of the eighteenth century in the study of nature based on evolutionary theory, genetics, ecological theory; but it is no accident that ecological theory—which is the basis of so much research in the study of plant and animal populations, conservation, preservation of nature, wildlife and land use management, and which has become the basic concept for a holistic view of nature—has behind it the long preoccupation in Western civilization with interpreting the nature of earthly environments, trying to see them as wholes, as manifestations of order."
Clarence J. Glacken, *Traces on the Rhodian Shore*

"Thoreau, like many other Romantics, was intensely interested in science, but he was less interested in the footprints of the Creator than he was in creation itself. He was disposed to find in nature not the result of some previous plan but a phenomenon continually expressive of creation; not the evidence of design but the design itself. Nature was the medium through which spirit manifested itself."
Robert Sattelmeyer, *The Natural History Essays*

FROM JULY 4, 1845, to September 6, 1847, during Henry David Thoreau's sojourn in a cabin near Walden Pond, he noticed something peculiar about people's perception of the pond. Most thought it was very deep. "There have been many stories about the bottom, or rather no bottom, of this pond, which certainly had no foundation for themselves. It is remarkable how long men will believe in the bottomlessness of a pond without taking the trouble to sound it," he wrote. "Many have believed that Walden reached quite through to the other side of the globe." And so he became interested in the depth of the pond and set out to learn this physical and quantitative characteristic of one of his favorite places in nature.

In his travels to the Maine woods, Thoreau had sought to understand nature and his relationship with it by making direct, careful, but informal natural history observations. But when these were not enough, he made use of the scientific method. The method begins with careful observations, expressed, if possible, in quantitative measurements. Back at home, Thoreau applied this more formal scientific approach to the question of the depth of Walden Pond. "Some who have lain flat on the ice for a long time," he wrote, "looking down through the illusive medium, perchance with watery eyes into the bargain, and driven to hasty conclusions by the fear of catching cold in their breasts, have seen vast holes 'into which a load of hay might be driven,' if there were any body to drive it, the undoubted source of the Styx and entrance to the Infernal Regions from these parts." There was a difference between Thoreau's approach and those of the people he described as "driven to hasty conclusions," which were based on conjecture rather than careful observation.

As a person with an intrinsic naturalist's and observer's inclination, Thoreau took a simple, direct approach to determining the depth of the pond: he measured it. He had the skill to do this because he worked from time to time as a surveyor. "As I was desirous to recover the long lost bottom of Walden Pond," he wrote, "I surveyed it carefully, before the ice broke up early in '46 with compass and chain and sounding line. I fathomed it easily with a cod-line and a stone weighing about a pound and a half, and could tell accurately when the stone left the bottom, by having to pull so much harder before the water got underneath to help me."

Thoreau made an important step from informal observation of natural history to quantitative measurement—a key step in using science to obtain a new kind of understanding of nature. Although this might seem to be an obvious route, it is not always the one taken. Just as people in Thoreau's time often avoided simple, direct measurement of natural phenomena, we, too, often prefer speculation and mystery to scientific study.

Once Thoreau had made one measurement, his curiosity was aroused, and he began to investigate the general shape of the pond's basin. He made more than one hundred measurements of the pond's depth. From these, he made a map, using his skills as a surveyor, and located the deepest point in the pond: "The greatest depth was exactly one hundred and two feet; to which may be added the five feet which it has risen since [with spring runoff into the pond], making one hundred and seven."

His curiosity further aroused, Thoreau began to consider generalizations arising from his quantitative measurements. "As I sounded through the ice I could determine the shape of the bottom with greater accuracy than is possible in surveying harbors which do not freeze over," he wrote.

Measurements led to surprises. "I was surprised at its general regularity," he wrote of the pond. "In the deepest part there are several acres more level than almost any field which is exposed to the sun, wind and plough. In one instance, on a line arbitrarily chosen, the depth did not vary more than one foot in thirty rods; and generally, near the middle, I could calculate the variation for each one hundred feet in any direction beforehand within three or four inches. Some are accustomed to speak of deep and dangerous holes even in quiet sandy ponds like this, but the effect of water under these circumstances is to level all inequalities."

Thoreau's investigation then progressed to ever more general theoretical constructs, leading him to develop a set of hypotheses about ponds and lakes in general. To do this, he had to find a means to aggregate his data so that he could see the results as a whole and think about that whole. For him, with his experience as a surveyor, this was the straightforward step of making a map, which required that his depth soundings be located geographically.

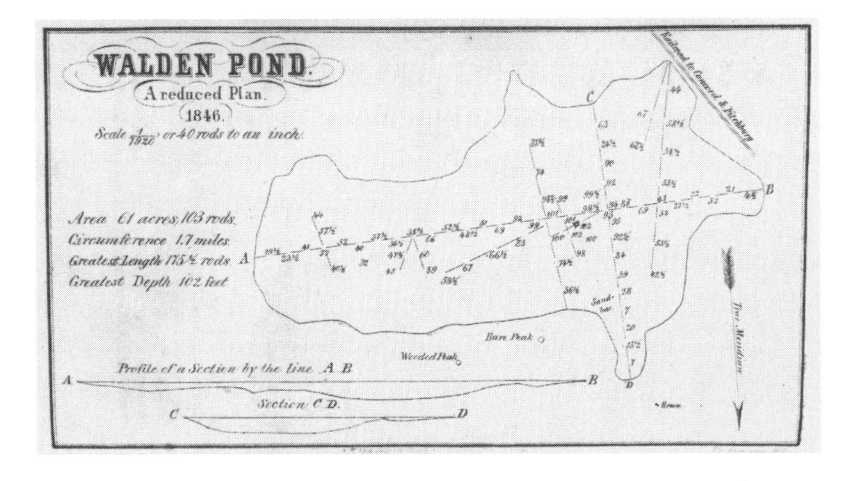

From the map he "observed a remarkable coincidence," he wrote: "the line of greatest length intersected the line of greatest breadth exactly at the point of greatest depth." Thoreau had expanded his inquiry beyond the initial question of the depth of the pond. Having made a series of measurements, he began to see the pond differently, as if its bottom were a field, and he became curious about the shape of that field. Making measurements had touched his imagination.

In reflecting on possible generalizations about his observations, Thoreau considered a comment made by somebody whose opinion he respected. "A factory owner hearing what depth I had found," he wrote, "thought that it could not be true, for, judging from his acquaintance with dams, sand would not lie at so steep an angle." In this process, Thoreau was, again, not the mythical hermit avoiding human contact but a person who considered the judgment of others when their experience and knowledge seemed valuable.

At this point, he was beginning to engage in an interesting thought process. Simple curiosity had led to a simple measurement and then to a series of those measurements, which in turn had led to a consideration of whether the measurements could be correct and, if so, what they implied. In this case, they implied that ponds could not always be shaped along the edges like dams of sand. "But the deepest ponds are not so deep in proportion to their area as most suppose," Thoreau continued, "and, if drained, would not leave very remarkable valleys. They are not like cups between the hills; for this one, which is so unusually deep for its area, appears in a vertical section through its centre not deeper than a shallow plate. Most ponds, emptied, would leave a meadow no more hollow than we frequently see."

On the basis of his series of quantitative measurements, Thoreau began to speculate about the shapes of ponds in general. He began to develop a hypothesis: perhaps in all ponds, the greatest depth tends to occur at the intersection of the line of greatest width and the line of greatest length. To test this idea, quantitative measurements were necessary. His scientific measurements had piqued new curiosity, led to new questions, while leading to a new understanding. The new understanding brought him, in a different way from before, closer to nature.

Thoreau's study of the pond brings out another important distinction, that between observations and inferences, which are ideas developed on the basis of a set of observations. A casual observation of Walden Pond's apparent depth at a given point is one thing; an inference that the entire pond must therefore be deep, made on the basis of that single glance, is another—it is a false inference. Confusing observations with inferences and accepting untested inferences is the kind of sloppy thinking often described by the phrase "thinking makes it so." Such sloppy thinking continues to pose problems in dealing with nature and the environment.

When scientists wish to test an inference, they convert it into a statement that can be disproved. This type of statement is known as a hypothesis. Once validated by tests and observations, a hypothesis continues to be accepted until it is disproved. A hypothesis that has not been disproved has not been proved to be true in an absolute sense; it has only been found to be probably true until, and unless, evidence to the contrary is found. In this way, science is a process; it is a way of knowing. It is not a set of fixed beliefs but a methodology by which beliefs can change over time.

Thoreau speculated more broadly about possible generalizations implied by his discovery that the deepest point of Walden Pond occurred at the intersection of the line of greatest length and the line of greatest width: "I said to myself, Who knows but this hint would conduct to the deepest part of the ocean as well as of a pond or puddle? Is not this the rule also for the height of mountains, regarded as the opposite of valleys? We know that a hill is not highest at its narrowest part." He began to speculate that he might "have elements enough to make a formula for all bodies of water."

To be scientific, generalizations such as this must be open to tests that could disprove them. If such a test cannot be devised, then a generalization cannot be treated as a scientific statement. Although new evidence can disprove existing scientific theories, science can never provide absolute proof of the truth of its theories.

Thoreau conceived of a test for his hypothesis: he made measurements of a second pond. "In order to see how nearly I could guess, with this experience, at the deepest point in a pond, by observing the outlines of its surface and the character of its shores alone, I made a plan of White Pond, a nearby pond, covering about forty-one acres," he wrote. He then located the line of greatest breadth and marked its intersection with the line of greatest length, the spot that he hoped would be the point of greatest depth. He then made enough soundings of this pond to determine the point of greatest depth, which he found to be "within one hundred feet" of his mark. "Of course, a stream running through, or an island in the pond, would make the problem much more complicated," he cautioned. Thoreau recognized that there were potentials for deviation from a specific theory so that a hypothesis might in general be "correct" but be subject to variation. Such deviation is what statisticians refer to as a source of variation or of one kind of "error."

Thoreau observed nature, his observations raised a question, and he answered that question with quantitative measurements, which in turn posed new questions and required additional measurements. Each set of quantitative measurements led to hypotheses that provoked him to make still more measurements and more generalizations, each open to testing. Thoreau instituted an iterative process that enhanced his scientific (i.e., outer) understanding of Walden and other ponds and, at the same time, must have altered his inner, spiritual relationship with them. Thoreau approached the study of ponds with the standard scientific method, which during his time was maturing in some fields of science and just beginning to be used in others, including the field of geology and the study of the relationships between living things and their environment.

This scientific method is usually considered to have its roots in the end of the sixteenth and beginning of the seventeenth centuries, with the work of William Gilbert (magnetism), Galileo (physics of motion), and William Harvey (circulation of blood). Unlike earlier classical natural philosophers who asked "Why?" in the sense of "For what purpose?" these scientists made important discoveries by asking "How?" in the sense of "How does it work?" Galileo also pioneered the use of numerical observation and mathematical models. The scientific method, which quickly proved very successful in advancing knowledge, was first described explicitly by Francis Bacon in 1620. Although not a practical scientist himself, Bacon recognized the importance of the scientific method, and his writings did much to promote scientific research.

It is not an exaggeration to say that in Thoreau's time, modern biology had not yet been born, whereas Sir Isaac Newton's laws of mechanics were well accepted and physics was a well-founded and maturing science. The study of natural sciences in Thoreau's time—geology, life sciences, and natural history—lagged behind the study of physics, astronomy, and chemistry.

Geology was in an early developmental stage during Thoreau's time, much as was biology, although one can make a good argument that geology was ahead of biology in most respects. Sir Charles Lyell, who lived from 1797 to 1875, published *Principles of Geology* in 1830–1833; his book is generally accepted as the first modern book on the science of geology. Lyell and his contemporaries had come to recognize that mountains are formed by uplifting generated by forces deep inside the Earth. Mountains and other areas that had been subjected to such uplifting were observed to be generally denuded of soils, which implied that they did not retain soils as they rose.

It was during Thoreau's lifetime that some geologists began to realize that glaciers had once covered vast areas of the northern continents. Jean Louis Rodolphe Agassiz developed at a young age into an active and famous scientist. In the 1830s, he discovered continental glaciation while roaming the Swiss hillsides and speaking with farmers. The farmers described large rocks and other elements of the landscape that Agassiz recognized as characteristic of land just to the south of existing glaciers. He therefore recognized that certain landforms created by mountain glaciers also occurred widely on flatter lands and could have been produced only by giant glaciers. It was becoming clear that the Earth's surface had undergone vast changes as a result of gargantuan geological forces. The landscape was not static, and it did not merely age; rather, it pulsed back and forth over long periods. Geologists in Thoreau's time were just beginning to understand these processes.

Some scientific progress had been made in biology by the early nineteenth century. The first step in any science—classifying, categorizing, and naming entities so that they can be discussed with clarity—was well under way. A science begins with the definition of static structures and, once these are understood, moves on to consider dynamic processes. This occurred with the development of the theory of biological evolution (a theory about processes), which was preceded by Swedish botanist Carolus Linnaeus's development of a formal method for naming organisms and classifying them according to similarities in static structural properties.

In the nineteenth century, that method for naming organisms and standardizing terminology provided a basis from which Charles Darwin, Alfred Russel Wallace, and others could begin a formalized consideration about the origin and evolution of species. The system for classifying plants and, subsequently, all forms of life—with a genus and species name in Latin—had been developed by Linnaeus in 1737 and was in widespread use by Thoreau's time. Thoreau used it himself. The Linnaean system ensured that individuals from different backgrounds and even speaking different languages, or different dialects of the same language, could know that they were speaking about the same thing. Informally, for instance, any tree that emits the pleasant scent of cedar is called a cedar tree. But the scent is emitted by many different conifers, some of which are not closely related, from *Chamaecyparis thyoides,* the southern white cedar of the eastern United States, to *Thuja occidentalis,* the northern cedar of the boreal forests of Maine, Canada, and west to Minnesota, to *Libocedrus decurrens,* the incense cedar of North America's Pacific Northwest, and numerous other tree species around the world. Before the Linnaean system was developed, two naturalists could discuss cedars and not realize they were talking about entirely different species. With the Linnaean system, the work of one naturalist was explicit to another, confusion was avoided, and biologists could build on one another's work.

The greatest idea of nineteenth-century biology, the theory of evolution of species, was developed during Thoreau's lifetime. It was not yet known when Thoreau built his cabin at Walden Pond in 1845, fourteen years before publication of Darwin's *On the Origin of Species.* Thoreau's trip on the Concord and Merrimack Rivers began on August 31, 1839; by that time, Darwin had completed his great trip on the *Beagle,* which lasted from 1831 to 1836, during which time he came to understand biological evolution. And Thoreau's first book, *A Week on the Concord and Merrimack Rivers,* was published in 1849, a decade before *On the Origin of Species.* More remarkable from our perspective, Thoreau's book about life at Walden Pond was published five years *before* Darwin's great work. It is remarkable because *Walden* seems to so many people a modern book, as if Thoreau had in his background the same scientific understanding of biological evolution as we do. Instead, the debate about how species originated was vigorous, and concern with the origin of species was in the air. Thoreau read Darwin's account of his voyage on the *Beagle,* his log of the journey that led him to develop the theory of evolution by natural selection, soon after its publication in 1839. He became acquainted with Darwin's *On the Origin of Species* in 1860, shortly after its publication.

Today, we view scientific research and religious inspiration as completely separate, but as Bradley P. Dean, editor of Thoreau's *Faith in a Seed,* wrote, "Thoreau's interest in science makes no sense apart from his Transcendentalist background." One of the beliefs of Transcendentalism was that the connection between people and nature had to be strengthened, and science—the results as well as the process of doing science—was inextricably linked to the strengthening of this relationship. Ralph Waldo Emerson, Thoreau's mentor regarding Transcendentalism, wrote that "nature is the opposite of the soul, answering to it part for part. One is seal and one is print. Its beauty is the beauty of his own mind. Its laws are the laws of his own mind," and "the ancient precept 'Know thyself,' and the modern precept 'Study nature,' become at last one maxim."

As Thoreau's study of the depth of ponds illustrates, science makes certain assumptions about the natural world: events in the natural world follow patterns that can be understood through careful observation and analysis, and these basic patterns are universal, holding throughout the universe now as in the past.

Although scientific progress is sometimes made through great leaps of insight that are then subjected to tests, most science progresses through a type of reasoning known as induction. The process begins with specific observations of the natural world that lead to generalizations, the same process Thoreau used in studying the depth of Walden Pond. Induction does not lead to absolute certainty; rather, it leads to a conclusion that is considered *usually* to be the case. When we say that something is "proven" by induction, we mean that it has a high degree of probability. But when we have a fairly high degree of confidence in our conclusions in science, we often forget to state the degree of certainty or uncertainty. Instead of saying, "There is a 99.9 percent probability that . . . ," we say, "It has been proven that. . . ." Such is our faith in science. Unfortunately, many people interpret the latter type of statement to mean that the conclusion is absolutely true, and this has led to much misunderstanding about science.

When all, or almost all, scientists agree on an observation, it is often called a fact. Such agreement requires repeated observations of the same kind and of the same phenomenon by many scientists. But this repetition is often lacking in the study of biological nature. The great complexity and diversity of life and the limited number of scientists working on these topics has precluded this. Generalizations are accepted on the basis of a single study by one scientist. With complex systems such as those of biological nature, lack of replication of observations becomes a major limitation on scientists' ability to generalize.

In mapping the basin of Walden Pond, Thoreau created what scientists call a model—something that is intentionally less than the entire reality but abstracts from that reality aspects that are of specific interest and limited enough in complexity to be studied. In science, models, which are a part of theory, are fundamental. A model may be an actual working model, a pictorial model, a mental model, a computer model, a laboratory model, or a mathematical model. Because hypotheses in science are continually being tested for consistency with nature and evaluated by other scientists, science has a built-in self-correcting feedback system.

As new knowledge accumulates, models may have to be revised or replaced, with the goal of developing ones that are more consistent with nature. Models that offer broad, fundamental explanations of many observations are called theories.

Although models are generally considered an integral part of the scientific method, the study of biological nature in the twentieth century had few models, and often those were never subjected to tests that could disprove them but were accepted anyway. This has happened repeatedly in the field of ecology and in other environmental sciences. In this way, the modern "scientific" study of nature often falls short of meeting the criteria of standard science. This makes Thoreau's maps of ponds all the more interesting: they were pictorial models that he subjected to testing.

As a whole, the ideas just discussed are usually referred to as the scientific method and can be summarized as a series of steps, such as Thoreau followed: Make observations and develop a question about the observations. Develop a tentative answer to the question, a hypothesis. Design an experiment to test the hypothesis. (Thoreau tried to do this by comparing the shape of Walden Pond with those of other ponds.) Collect data and compile them in an organized manner, as in a table. (Thoreau's data and his map follow this step.) Interpret the data through graphic or other means. Draw a conclusion from the data. Compare the conclusion with the hypothesis and determine whether the results support or disprove the hypothesis, as Thoreau did in his attempt to extend his theory about the shape of a pond basin from Walden Pond to another pond. If the hypothesis is accepted, conduct further tests to support it. If the hypothesis is rejected, make additional observations and construct a new hypothesis.

"Now we know only a few laws," Thoreau wrote in his discussion about the depth of Walden Pond. He was reflecting on his ideas about the shapes of ponds, and from this reflection he was beginning to think about the character of nature in a fundamental way: whether nature was everywhere harmonious and balanced in structure, as it seemed to be in the ponds he was studying. "Our notions of law and harmony are commonly confined to those instances which we detect; but the harmony which results from a far greater number of seemingly conflicting, but really concurring, laws, which we have not detected, is still more wonderful," he continued. "The particular laws are as our points of view, as, to the traveler, a mountain outline varies with every step, and it has an infinite number of profiles, though absolutely but one form. Even when cleft or bored through it is not comprehended in its entireness."

Here, Thoreau touched on an important aspect of science: the search for patterns and regularities and an explanation for them. The conception of possible patterns and regularities is limited by the scientist's imagination and is affected by the general ideas and outlooks of the culture of the time.

A common misunderstanding about science arises from confusion between the use of the words *theory* and *law* in science and their use in everyday language. A scientific law is a generalization that is supported by a large number of observations and tests and describes orderliness of some kind about things and events and their changes over time and space. A scientific theory is a grand scheme that relates and explains many observations; either it is a set of explicitly stated conjectures, as in Albert Einstein's mathematical theory of relativity, or it is supported by observations.

In contrast, in everyday language, a theory can be a guess, a hypothesis, a prediction, a notion, or a belief. The phrase "It's just a theory" may make sense in everyday language but not in the language of science. Scientific theories generally have tremendous prestige and are considered the greatest achievements of science. The term *scientific law* is usually restricted to major generalizations that have many implications and are unvarying, such as the second law of thermodynamics or Newton's laws of motion. Thoreau's measurements of the depth of Walden Pond expanded into an insight regarding possible regularity in the shapes of naturally occurring water basins, which led him to discuss the general status of scientific laws in his time.

Science is limited by the technology available. Mapping of vegetation, for example, became much easier with the invention of aerial photography. Today, very complex lake basins can be mapped with sonar devices, providing much more complete maps than Thoreau was able to make. Color film makes maps easier to use. The invention of satellite remote sensing, digital recording, and small computers has made it possible to make maps based on quantitative information much more efficiently and for much larger areas than before. New computer-based information-processing systems, especially geographic information systems, allow investigations of relationships among many factors that were difficult or impossible to carry out before. Scientific tools and methods such as these form a basis for new science, but they are also products of previous science. Their development demonstrates one way in which science is a process, not a fixed set of beliefs.

Thoreau's study of the depth of Walden Pond followed the classical scientific method. Scientists generally agree that in many cases, this is an idealized process. In studying complex systems, including biological systems, the scientific method cannot always be applied in this standard way. Ecology is such a new science, in terms of the development of general laws and theories and the collection of valid data, that in many cases it remains unclear what the best approach would be. Quite often, as a result, casual and poorly informed observations form the basis of policy. It is common with environmental issues to set policy by plausibility. By this, I mean that government agencies decide that if it looks good, it must be true. As a result, one surprising thing about current environmental issues is the common failure to use the scientific method—to follow the process that Thoreau followed more than one hundred years ago in his study of the depth of Walden Pond.

A classic example of this occurred in the 1940s in the state of Oregon. People noticed that in drought years, adult salmon had a difficult time swimming upstream over large logs, which are common in streams of the Pacific Northwest. The wood, still in good condition, was valuable in the World War II effort—cedar was used to encase batteries for battle tanks. As a result, a statewide policy was set in motion to remove large logs from salmon streams. The result was disastrous to salmon because these woody debris dams created systems of pools and fast-moving water, both of which were essential to the salmon. In this case, action was taken without careful tests first being conducted. Policy makers accepted as truth casual observations made by a few individuals, in contrast with Thoreau's insistence on testing the validity of an idea for himself.

There are several aspects of modern science that we sometimes confuse and therefore need to distinguish: science as a way of knowing; the effects on the scientist of the experience of doing science; science as a set of findings, which are often interpreted as a fixed set of beliefs to be taken up permanently by others; and the influence of scientific discoveries on our imagination, emotions, and sense of place in the universe—our spiritual and religious feelings and ideas. Thoreau's experiences in measuring the depth of Walden Pond touch on all of these.

I have just outlined the standard description of the scientific method. However, in real life, scientific discoveries are more complex. Frequently, having detailed knowledge of a phenomenon makes one alert to the unexpected. As a result, accidental discoveries—serendipity, being at the right place at the right time—and flashes of insight and creativity frequently play a role. Often, a spark of imagination, or a leap of insight, as it is commonly put, may arise from contemplation of a few informal observations made by others.

The great mathematician Bertrand Russell described an analogous process in his work. He wrote that he would concentrate on a given question or problem as hard as he could for a number of days, until he could think about it no more. Then he would put it aside. Some days later, the answer would "appear" to him, seemingly out of the blue. This suggests that there are subconscious mental processes linked to human reason and insight.

Serendipity, too, is in reality more than being at the right place at the right time. The discovery of penicillin may appear to have been a product of that kind of luck—an accidental observation that bacteria did not grow near bread mold—but it took an experienced bacteriologist to make that observation. In 1928, Sir Alexander Fleming noticed that a bacterium that caused infection, *Staphylococcus aureus,* did not grow in a laboratory culture that had been accidentally contaminated by a common bread mold, *Penicillium notatum.* Bread mold has certainly been a familiar object in the history of civilization, and there must have been many opportunities to observe its effects, if one only knew what to look for. Yet it took an expert on those forms of life, who had thought much about them, to realize that a spot in a culture barren of bacteria could be connected to the growth of bread mold. Fleming's detailed knowledge and background were essential to his "being at the right place at the right time." This illustrates the value of detailed observations of nature, which Thoreau emphasized so strongly for himself.

Although he did not discuss the hardships of measuring the pond's depth through layers of ice, Thoreau appreciated the difficulty of scientific observations, an appreciation very likely stimulated by his own experiences. He wrote of other scientists before him: "What an admirable training is science for the more active warfare of life! Indeed, the unchallenged bravery which these studies imply, is far more impressive than the trumpeted valor of the warrior." Specific examples of the hard effort of others came to his mind from readings of the classics. "I am pleased to learn that Thales was up and stirring by night not unfrequently, as his astronomical discoveries prove," he continued. Of more recent experts, he wrote: "Linnaeus, setting out for Lapland, surveyed his 'comb' and 'spare shirt,' 'leathern breeches' and 'gauze cap to keep off gnats,' with as much complacency as Bonaparte [viewed] a park of artillery for the Russian campaign. The quiet bravery of the man is admirable. His eye is to take in fish, flower, and bird, quadruped and biped."

From this, Thoreau speculated: "Science is always brave; for to know is to know good; doubt and danger quail before her eye. What the coward overlooks in his hurry, she calmly scrutinizes, breaking ground like a pioneer for the array of arts that follow in her train. But cowardice is unscientific; for there cannot be a science of ignorance. There may be a science of bravery, for that advances; but a retreat is rarely well conducted; if it is, then is it an orderly advance in the face of circumstances."

Imagine Thoreau out on the pond ice in winter, making one measurement after another. The more he made, the more questions occurred to him and the more measurements he was led to make. Picture yourself doing this—standing or sometimes lying on your belly on the ice, trying to get a line down a hole; making sure that the rock at the end of the line is touching the bottom; writing a number in a notebook with wet hands numbed by the ice and wind; protecting the paper; hauling up the rope as it sprays cold water on your clothes; reading the numbers and thinking about what to do next. This is "contact, contact, contact" with nature, driven not by a quest for the sublime or for recreational adventure but by a scientific search for knowledge.

This leads to another question that occupied Thoreau: whether use of the scientific method is helpful or destructive to nature's spiritual and creative effects on people. What is the role of science, as a study of physical, tangible phenomena, in our attempt to make spiritual contact with nature? Is science an aid or an obstacle—a pathway or a closed door?

Thoreau was active in science throughout his life, although he was not a major scientific figure of his time. In spite of his persistent, lifelong practice of science, he repeatedly fell victim to self-doubt, questioning the utility of science in his search for a spiritual connection to nature. His quest to apply science and his repeated questioning of the effect of scientific activities on his intimate, intangible contact with nature is a dilemma we all face, though we rarely confront it as directly as did Thoreau. His struggle is therefore instructive to us today and can be seen as a metaphor for our dilemma over the same issue.

In my own experience, I can attest to the validity of science as a path for making contact with the intangible in nature. When I was a graduate student, I had the good fortune to be a teaching assistant for Professor Murray Buell, a well-known plant ecologist of the mid-twentieth century. Murray was the same kind of intrinsic naturalist and scientist as was Thoreau, a person continually fascinated by nature, not simply seeking to observe it. He was driven to study and measure from a great, deep love of nature.

One day, he and I visited a bog—much like the one in which Thoreau was lost in Maine, but smaller. We wanted to make sure that a device for sampling the mud from the bottom of the pond, a corer, would function properly when we brought students to the bog on a field trip. It was a beautiful spring day, with a gentle breeze atomizing the aromas of soil, water, breaking buds, and spring flowers. We walked a mile or so toward the bog through pleasant woodlands of oak and hickory, then onto wetter ground, and finally onto the surface of a quaking bog. A quaking bog is called that because it is a mat of soil and vegetation that floats on the surface of open water. It is strong enough to serve as an anchor for trees and to hold up a person. But if you jump up and down on one, it quakes. The thinner ones roll like a ship at sea when you walk across them.

The geological and ecological genesis of a quaking bog is well understood today. During the past ice ages, which came and went between hundreds of thousands of years ago and about ten thousand years ago, the continental ice sheets bulldozed the soil. Sometimes large rocks would become lodged in the bottom of the ice and dig through the soil like the teeth on a bulldozer blade. As the ice melted, large basins were left scattered on the landscape. Occasionally a large chunk of ice, a landlocked iceberg, would dislodge and form a dam. Water flowing from the melting glacier would form a pond or lake in the depressions behind the ice or debris dams.

As the climate warmed, various kinds of vegetation returned, including sedges adapted to such climatic variations. Sedges are typically the first flora of a quaking bog. They put floating runners out on the surface of the water, and new sedge stems grow from the runners and form a mat. Wind-borne dust falls onto the mat, beginning a layer of soil. As the older sedge stems and runners die, they add organic matter to the floating soil. Eventually, the soil becomes hospitable to seeds of other plants that are blown onto the mat or dropped there by birds and small mammals. Shrubs and trees that can withstand having wet roots in acidic waters, such as cranberries and their relatives and southern white cedar, germinate and grow.

This is a process known as bog succession—the succession of plants that occupy and fill in a bog, bringing terrestrial life where there was none before. The floating mat of sedges plays a crucial role in reestablishing biological nature after the great catastrophe of a glaciation and after smaller disturbances that alter a landscape enough to create open water. Over time, the open water gradually fills in. Beginning near the edges of the pond or lake, waters flowing in also deposit sediments on the bottom. The pond or lake begins to fill in from both the top and bottom. Eventually, the floating and subsurface deposits meet and the soil no longer floats. As an organic mat builds up, its surface becomes drier. Eventually, these raised areas become high enough to support trees that cannot grow in very wet soils. Thus, after continental glaciation, bogs fill in slowly from the edges. The area of open water decreases, sometimes vanishing completely.

I knew this geological and ecological history intellectually, as something abstract that had occurred thousands of years ago. But on this day, Murray Buell and I were striding over just such a bog, bouncing on its quaking surface. Although I had visited quaking bogs before, this was the first time I had done so as a student of ecology, accompanied by one of the country's experts in plant ecology.

When we reached a suitable location out on the bog, we jumped up and down a few times for fun and to make sure we were really over open water. Then we stopped and assembled the corer. This is a device like a long metal fishing rod that one can carry in sections into the woods. The sections have threaded ends and can be put together to form a pole forty or fifty feet long or more. At the leading end is a tube constructed so that it can be forced into the soil to any depth and then, with a twist, locked closed so that the observer, reading off the length that the pole has descended into the soil, knows the exact depth of the sample. We threaded the device together and pushed it into the soil. The soil resisted. We pushed and twisted and threaded the corer downward: five feet, ten feet, then into the water below, through which the corer moved rapidly until it abruptly struck a solid rock bottom. The steel rods clanked and transmitted the impact directly to us, to our ears in a sound and to our hands in a vibration. The corer's pipe was down fifty feet, the equivalent of a five-story building. Standing on the floating mat, suspended above the water, we could pull the rod up and down and feel the solid rock far below.

With that clanking sound, I had an entirely different idea of where I was. I was fifty feet above the bottom of a lake whose base was hard rock, standing as if I could walk on water. Looking around, I saw the shape of the basin on whose top we stood. Thousands of years of geological events, which I had only read about, suddenly became real: ten thousand years ago, mile-high sheets of ice had gouged the rock, digging a bowl in the once level land. In my imagination, I saw two landscapes: the present one and an ancient one. The ancient one formed a setting and surrounding for the modern. I could imagine the bare sand gouged out by the ice, empty of vegetation, existing in a cold climate and forming a bowl like the shape of Walden Pond. Geological and ecological processes became real. I felt a different kind of connection to the landscape, a connection that extended back in time. Suddenly, I saw landscape development as an active process. Where we stood, there was both water and soil beneath our feet. I could both understand intellectually and feel viscerally two of the unique features of our planet: life and water (in its three phases, gas, liquid, and solid). These had created the topography on which we stood, a living, fluid, dynamic setting that would appear to one not standing where we were as still and static as a painting.

Doing science had transformed my feelings. It was a creative moment in the sense that it enabled me to discover my connection to nature, present and past. Here was a new world, my understanding of which was made possible by more than a century of observation and theories about glaciation, begun by Agassiz and made immediate by the *thunk, thunk, thunk* of steel on granite. Only a moment before, we had been simply struggling through a wet woods. Our use of quantitative methods unknown during Thoreau's time created for me a new sense of nature, which I have never forgotten. This experience comes back to me when people question how one can study nature and appreciate it at the same time, or how a scientist can have the same emotional feelings about nature as someone else who believes he alone truly loves his environment.

This sense of the vividness of nature not just now but also long ago became real for me in another experience at another site, an ancient glacial lake in New Haven, Connecticut, that had filled with sediment. The clay from this pond, formed when inflowing streams slowed as they entered it and dropped the heavier particles of sand and silt they carried, had been mined to make bricks and pottery. Some scientists at Yale University, where I was on the faculty at the time, had studied the clay basin to learn the history of the vegetation in this part of Connecticut. One day, several of us went out to visit this clay pit, which was at the edge of a dismal industrial zone of New Haven. The large hole in the ground was hardly inspiring, but there at the bottom of it, a log of a butternut tree had been found intact. The clay just above the level where the log was exposed had been deposited about six thousand years ago. Secure in an oxygenless, acidic water environment, the log could not decay. The wood looked as fresh as if it had fallen into a lake a few days before. Tom Siccama, a colleague of mine at Yale, cut the log in half with a chain saw so that we could examine the growth rings. The wood was solid to the core.

Butternut trees grow on floodplains; the log had probably floated down a stream and into a lake. Seeing the log, I suddenly imagined that ancient scenery: an open lake with a stream flowing into it along a rolling countryside. Perhaps Indians strolling by had seen the log float downstream. The lake might have been surrounded by other butternut trees. Contact with a six-thousand-year-old landscape was made fresh to us on that day by a combination of new scientific methods and discoveries: the ability to date buried materials using carbon 14, a twentieth-century invention; natural history knowledge of the habitats that favored butternut; and an understanding of the geological processes that led to the formation and filling of the bog. The physical awareness of nature made possible by technology and scientific understanding not available before our century changed my sense of contact with my surroundings.

It seems to me that Thoreau similarly grasped a new appreciation of the topography beneath Walden Pond when he became engrossed in measuring its depth and mapping its basin. The process had involved aesthetic as well as scientific appreciation. "The regularity of the bottom and its conformity to the shores and the range of the neighboring hills were so perfect," he wrote, "that a distant promontory betrayed itself in the soundings quite across the pond, and its direction could be determined by observing the opposite shore. Cape becomes bar, and plain shoal, and valley and gorge deep water and channel." The landscape beneath the pond had become visible in his imagination. The process of doing science had affected Thoreau's imagination, emotions, and sense of his spiritual place in the universe, just as did the results, the scientific findings.

By Thoreau's time, the results of science—scientific information—had greatly affected people's ideas of beauty and the ways in which they responded to nature. As an example, before the nineteenth century, mountain scenery was not considered beautiful or a place for spiritual inspiration in Western civilization. Since Greek and Roman times, symmetry had been the key to beauty, in nature and in artifice. Mountains were the warts and wrinkles on an aging Mother Nature's skin. Religious inspiration was to be found in the neatness of natural symmetry. Only with the increasing technological power of the emerging scientific-industrial era did people become free enough of their surroundings, powerful enough in relation to nature, to be able to pass through the Alps in Europe simply for recreation or travel the woods of Maine to enjoy the outdoors, thus experiencing nature in a different way.

Also underlying this shift in people's idea of natural beauty was a change in perception of the cosmos. With the work of Galileo, Johannes Kepler, Nicolaus Copernicus, and Newton, a transition had occurred from belief in an Earth-centered, static universe to awareness of a cosmos of motion and process. And with the acceptance of the laws of the dynamics of planets and stars, motion and power came to be appreciated as aesthetically pleasing, in a different way from the prettiness of quiet, symmetrical formal gardens. The power of ocean waves, previously seen only as terrifying, became a reflection of the power and magnificence of God and, therefore, a representation of beauty. The ragged, steep, snowy Alps, previously perceived as horrible, became awesome. This transition took place as the eighteenth century ended and the nineteenth century began.

Meriwether Lewis touched on these aspects of natural beauty in describing his arrival at the five great waterfalls of the Missouri River, just downstream from the present location of Great Falls, Montana, in June 1805. He saw one beautiful falls on June 13 and another the next day. The latter, which he called Rainbow Falls, is now much altered by Rainbow Dam. It was, he wrote, "one of the most beautiful objects in nature." Lewis spent hours trying to decide which of the two, the falls he had seen the day before or this one, was the most beautiful. "At length I determined between these two," he wrote, that Rainbow Falls was "pleasingly beautifull," whereas the one he had seen the day before was "sublimely grand." These turns of phrase were in use among the Romantic poets and their predecessors to describe aspects of beauty. The word *beauty* was used then in the classical Greek and Roman sense of symmetry, perfection in geometry—the old idea. *Sublime,* on the other hand, had come into fashion among the Romantic poets in reference to the awe-inspiring scenery of the great mountains in the Alps—the new idea, arising from a dynamic perception of nature. It is against this backdrop of aesthetics, philosophy, and religion, as well as the development of the sciences, that Thoreau viewed Mount Katahdin, Umbazooksus Swamp, Walden Pond, and the woods around Concord.

Natural grandeur, as popularized by the Romantic poets, was savage but powerful; it was *awe*-ful, both in the sense of the derivation of that word, inspiring a feeling of awe in the onlooker, and in the modern connotation, frightening, awful. This linkage of Thoreau's view with that of the Romantic poets was perhaps more than accidental, for Thoreau was an avid reader and kept abreast of the major works of European and American civilization. In many ways, his approach to nature was consistent with those of William Wordsworth and his colleagues. But this conception of nature, familiar and perhaps in some ways commonplace to us, was, at the end of the eighteenth century and the beginning of the nineteenth, revolutionary.

The religious debate in Thoreau's time about the theory of evolution illustrates how changes in scientific understanding affect people's religious beliefs, just as Newton's discovery of the laws of motion and Galileo's and Kepler's discoveries had done in earlier times. In both cases, new scientific information appeared to deny an aspect of the structural perfection of the universe. And in both cases, belief in structural perfection was replaced by a recognition of elegant, beautiful ideas about processes. The gradual acceptance of new scientific knowledge by Western religions can be viewed as a replacement of the ideal of structural perfection with an idea of perfection in processes; dynamics replaced stasis.

Our cultural heritage, *including science as part of the heritage,* affects our ways of thinking and feeling about nature. At various periods in his life, Thoreau saw science as a clear path to both nature-knowledge and nature-contact, to both the tangible and intangible. "The true man of science will know nature better by his finer organization," he wrote; "he will smell, taste, see, hear, feel, better than other men. His will be a deeper and finer experience. We do not learn by inference and deduction and the application of mathematics to philosophy, but by direct intercourse and sympathy."

Thoreau never quite resolved the conflict in his mind about whether the process of doing science aided or hindered his ability to make spiritual and creative contact with nature. For myself, my experiences led to a clear answer: the search for the tangible increased my sense of contact with the intangible. The sound of a steel rod against granite as I stood on the floating mat of a bog changed forever my view of nature and my inner sense of connection with it. It was an important moment for me. Our society struggles to find a confluence of the seemingly separate paths of rational, scientific knowledge and an inner appreciation of nature. Thoreau's travels, natural history observations, and study of the depth of Walden Pond may be keys to unlock the gates that bar access to such a confluence.

The crudeness of modern ecology as a science, and the lack of clarity as to what is the best possible approach in many cases, opens ecology to a kind of faulty decision process. It is one of the factors that cloud the distinction between science and nonscience in debates about nature. These debates are much like the speculations about the depth of Walden Pond that Thoreau criticized, based on conjecture rather than careful observation. When data are obtained, they are often incomplete, or the methods used change over time so that measurements made at one time cannot be compared with previous measurements. Sometimes the incompleteness involves a single factor; for example, in measurement of the depth of a pond, the number of measurements may be inadequate or, if a sequence of measurements over time is required, the sequence is broken. This has been the case in forestry, limiting scientists' ability to make sound statements about what is required for sustainable forest practices.

Similarly, I recently sought to find out the effects of pesticides on fish in the Missouri River. As one of North America's major rivers, the Missouri ought to be well studied. Given that it drains one-sixth of the United States and much of this fraction is agricultural land, I thought measurements of the effects of agricultural runoff would be easy to obtain.

Before channelization of the Missouri River, measurements of water quality were few and scattered, but these measurements do provide some baseline data. These suggest that by 1984, nitrate and phosphate concentrations in the lower Missouri River—such as downstream from Kansas City—had increased to four times the baseline level. Meanwhile, upstream in the reservoirs, nitrate and phosphate levels seem to have decreased.

Harmful side effects of agriculture first became apparent in 1964, when a fish kill extended more than 100 miles downstream from Kansas City, Missouri. Even so, monitoring of pesticides remained spotty, but between 1968 and 1976, studies of fish flesh at Council Bluffs, Iowa, revealed concentrations of the pesticide Dieldrin that exceeded public health standards in 13 percent of the samples, and concentrations of DDT and its breakdown products exceeded public health standards in one-third of the samples. In the early 1970s, concentrations of Dieldrin, Aldrin, and polychlorinated biphenyls (PCBs) in fish analyzed at Hermann, Missouri, posed a potential health threat. In the mid-1970s, Dieldrin levels were high enough in catfish that the Missouri Department of Conservation issued warnings. People stopped buying catfish, and this affected commercial fisheries.

Today, about 700 million pounds of more than 100 pesticide compounds are applied each year in the United States, and herbicides account for about 60 percent of the total pesticides found in the country's waters. Public health standards and standards for environmental effects have been established for some but not all of these compounds. As a step in compiling information for my book *Passage of Discovery,* about the Missouri River and how it has changed since the Lewis and Clark expedition, I searched the scientific literature and the World Wide Web and called many of my scientific colleagues who study rivers and organic chemicals as well as scientists I had met while exploring the Missouri River countryside. Surely, I thought, scientists had conducted experiments to determine what happens to pesticides applied to row crops—how fast the chemicals decay, how fast they are transported to nearby rivers by water flowing on and below the surface. Surely experiments had been carried out to determine what scientists call the "dose–response curve"—the curve on a graph showing effects on a species as the concentration of a toxin increases. But I could not find such studies for any location in the Missouri River valley or for fish and wildlife species found there. All my colleagues agreed that the research needed to determine the effects of these chemicals on Missouri River fish and wildlife had not been done.

Some things were being done. Monitoring of concentrations of these chemicals in the waters had increased greatly. The U.S. Geological Survey had established a network to monitor sixty watersheds across the United States. One of these is the watershed for the Platte River, one of the Missouri's major tributaries. The most common herbicides used in growing corn, sorghum, and soybeans along the Platte River were Alachlor, Atrazine, Cyanazine, and Metolachlor, all of them organonitrogen herbicides. Monitoring on the Platte near Lincoln, Nebraska, suggested that during heavy spring runoff, concentrations of some herbicides might reach or exceed established public health standards. But research of this kind continues to be sparse, and as a result it is difficult to reach definitive conclusions—or even reasonable approximations of conclusions—about whether concentrations are endangering public water supplies or harming wildlife, fish, freshwater algae, or vegetation. The advances in knowledge tell us on a more regular basis the concentration of many artificial compounds in sampled waters, but we are still uncertain of their environmental effects.

The search for adequate information about chemicals in the Missouri River was frustrating. Here was a potentially major national problem that had received little attention; most of the focus on pollution has been on urban and industrial pollution and pollution from feedlots, much less of it on pollution from row crop agriculture. By introducing large quantities of artificial compounds into our rivers, we are de facto conducting experiments on nature without the usual features of the scientific method: treatments and controls, adequate monitoring, and adequate long-term experimentation in laboratories.

There are two possible reasons for our failure to use the standard scientific method in approaching problems such as the effects of pollutants in rivers. The first is that the standard scientific method would work in theory, but it would require the collection of a huge volume of data that would be impractical to obtain. Especially difficult would be making measurements at the right times in the right places and in the proper relationship to one another in order to obtain a well-organized body of information, like Thoreau's map of Walden Pond.

The second possible reason for our failure to use the standard scientific method is that the complexity of nature's ecosystems makes them intractable to that method. At present, there is a vigorous debate going on among scientists about the characteristics of complex systems and whether a new kind of scientific method has to be developed for these systems. Whatever the outcome of this debate, the common principles of the scientific method remain: to begin with accurate and objectively made observations and to try to find ways to generalize from these. Perhaps a different way of seeking generalizations will be part of a new scientific method. The generalizations that result must, of course, be open to testing that could disprove them.

The use of science to study nature raises many questions. When there is a claim that it has been used, why does it so often seem unsuccessful in solving environmental problems in our times? As I have suggested, often it is because the standard scientific method either is not used or is used incompletely or incorrectly. Moreover, decisions often are desired or even required by law before adequate scientific information becomes available. Once again, Thoreau serves as an interesting and useful guide.

In part, we fail to use science appropriately in dealing with environmental issues because we impose values on knowledge. There is a tendency to begin with an ideology—an idea not supported by facts, sometimes based on intangible beliefs. There is a distinction here between starting with an ideology—an assumption about how nature *must* work—and starting with a hypothesis—an idea proposed as possibly correct but open (both in one's mind and in reality) to tests of its validity, tests that have the potential to disprove it. If we start with an ideology, we will go wrong. We will have closed ourselves off from real knowledge of the physical qualities of nature. Once again, the distinction becomes clear between science as a way of knowing and science taken as a set of fixed beliefs. If we start with a hypothesis, we have a chance of being right, and even if we are wrong, we have an opportunity to learn more about nature.

An irony of our scientific age is that we believe in science but often fail to use it when it comes to our connection with nature. Instead, we tend to proceed from ideologies often based on hopes, desires, and spiritual and religious beliefs that are not subjected to scientific tests. We make policy by plausibility. Although we believe that the science of nature is separate from our intangible connections with nature—our sense of joy, creativity, and inspiration—this is not the case.

Thoreau, in contrast, was able to make correct use of science while appreciating both the spiritual and physical qualities of nature. He was able to separate these as concepts but combine them in activities. Thoreau understood science as a way of knowing, and he understood its value in comparison with other paths to different kinds of knowledge. The scientific method was an important process to Thoreau, one he followed throughout his life. He could not, of course, study everything as carefully as he did the depth of Walden Pond. There were many topics not open to him for direct study and topics with which he was not familiar. In such cases, he needed the expertise of others. Sometimes he found this expertise in people he met; at other times he found it in their writings, sometimes in the great writings of past civilizations.

Chapter 8

The Poet and the Pencil

"The machine does not isolate man from the great problems of nature but plunges him more deeply into them."

Antoine de Saint-Exupéry, *Wind, Sand, and Stars*

"The speed with which Science marches from discovery to discovery, leaving behind her a trail of shattered theories, compels the most dogmatic of us to be cautious."

Henry C. Kittredge, *Cape Cod*

IN 1834, Henry David Thoreau made a trip with his father to sell pencils in New York City. Pencil making was the family business, and apparently the sales trip was needed to provide for the young man's schooling. Several years later, having finished his schooling, Thoreau set about to try to improve the pencil.

The pencil, first made of a thin rod of lead in a wooden case, was an invention of the Renaissance. But in the seventeenth century, graphite replaced lead. Graphite, sometimes called "black lead" and the "lead" of a modern pencil, is a form of carbon that occurs as a mineral. The first graphite pencils were made of thin rods of graphite glued inside a wooden case. This required purity of the graphite produced by the mines; impurities, such as small grains of silicon dioxide, a common form of sand, would make the pencils write roughly. The invention of the pencil can be seen as the culmination of the search for a simple writing tool whose marks were clear but could be erased. The first graphite pencils were made in Germany and Great Britain, but the best of these originated in a mine at

Borrowdale, a valley near Keswick in northwestern England, which yielded the purest mineral. Over time, Borrowdale graphite became rare, and the English government imposed restrictions on its sale that made it difficult to obtain for pencil manufacturing outside the British Isles.

Graphite found elsewhere had many more impurities and made poorer pencils. Graphite occurred as a mineral in New England, but its quality was not as high as that of Borrowdale. Thoreau, active in the family business, used his inventiveness to improve pencils made with the poorer-quality New England graphite. Previously, Thoreau's father had ground graphite to remove impurities, but the process was not perfect; the graphite in his pencils was often gritty, and the pencils did not write smoothly. Independently, Thoreau came up with the idea of mixing graphite with clay and then firing the mixture in a kiln to make a solid rod, a process that had been discovered in Europe but was not in use in the United States at the time. Varying the percentages of clay and graphite altered what we refer to today as the hardness of the lead.

Thoreau worked out all the mechanical details of a new purification process himself. First, the graphite was ground to a fine powder in a water-powered mill; then a water-powered fan blew the powder upward. Carbon is a relatively light element, and the impurities, such as sand, tended to be heavier and could not reach the same height as the graphite powder. Blowing the powder upward separated the various particles by molecular weight. The purified graphite, blown the highest of all the materials in the ore, was then captured in a cloth container. One of Thoreau's close friends, the son of Ralph Waldo Emerson, Thoreau's mentor, described this graphite-separating machine as a "narrow churn-like chamber around the mill-stones prolonged some seven feet high, opening into a broad, closed, flat box, a sort of shelf. Only the graphite powder that was fine enough to rise to that height, carried up by an upward draught of air, and lodged in the box was used, and the rest ground over."

Thus, a new picture begins to emerge of Thoreau as a young man—not only a naturalist, surveyor, and occasional lecturer and devoted writer but also an inventive participant in his family's business.

Thoreau began keeping a journal on October 22, 1837, but according to Henry Petroski in his fascinating book *The Pencil,* "over the following decade, during which time he was engaged on and off in the business, he would mention pencil making rarely and then only in passing." In 1841, at

the age of twenty-four, Thoreau moved into the Emerson household. He stayed there until 1843, returning to work at his father's pencil business when he needed money or when the business needed his help, again something about which Thoreau did not write.

The pencil business was far from a passing interest to Thoreau. After living with the Emersons, Thoreau went to Staten Island for eight months, taking a position as a tutor. During that time, he often wrote to his father, asking about the pencil business. After he returned home, Thoreau rejoined the family business. He continued to apply his engineering inventiveness to the pencil, developing new ways to fit the graphite within the wood casing. "He apparently conceived of many ways to improve still further the processes and products of the factory, and according to Emerson could think of nothing else for a while," Petroski wrote.

Ralph Waldo Emerson was well aware of the Thoreau family's pencil business and Thoreau's contributions to it, as indicated in a letter he wrote to a friend, Caroline Sturgis, dated May 19, 1844:

> Dear Caroline
>
> [I] only write now to send you four pencils with different marks which I am very desirous that you should try as drawing pencils & find to be good. Henry Thoreau has made, as he thinks, great improvements in the manufacture, and believes he makes as good a pencil as the good English drawing pencil. You must tell me whether they be or not. They are for sale at Miss Peabody's as I believed, for 74 cents the dozen. . . .
>
> Farewell.
> Waldo

Sturgis wrote back to Emerson: "The pencils are excellent—worthy of Concord art & artists and indeed one of the best productions I ever saw from there—something substantial & useful about it. I shall certainly recommend them to all my friends."

Thoreau wrote little about this aspect of his life. Petroski wrote that this is typical of engineers, who tend not to write about their inventions; instead they make drawings and work from their imagination directly. But it does seem surprising that one of America's literary giants would rarely mention in his own writings an activity that occupied him on many occasions.

Petroski concluded, "There is little doubt that before Henry David Thoreau was the literary celebrity he has come to be, the pencils he and his father made came to be without peer in this country," and "thus, contrary to the conventional wisdom then and still current around Concord and elsewhere, Henry David Thoreau was no slouch, even though in May 1845 he left home and the pencil business and began to build his cabin near Walden Pond where he would live until 1847." Petroski noted rather wryly that Thoreau continued a kind of chemical engineering at Walden in the form of creative bread making, apparently inventing raisin bread, which was "said to have shocked the housewives of Concord."

Petroski summarized Thoreau, his reputation, and his involvement with the pencil as follows:

> In the Concord, Massachusetts, Free Public Library there are shelves of editions of Thoreau's *Walden* and shelves of books on the author's times, writings, and thoughts. One catalogue of the Thoreau Society's archives lists more than one thousand items, but the number of those dealing specifically with Thoreau as pencil maker and engineer of pencil-making machinery is nil. While a "nail picked up at the Thoreau cabin site" is included among the literary works, no pencil is. The collection includes a single Thoreau & Company pencil label (reprinted in ink of course), and that is the one artifact that gives any hint of the activity that provided the family income. One must learn of Thoreau the pencil engineer almost by inference from the few scanty references within more general works that the curator happens to recall. There are now a few pencils among the books and literary material in the Thoreau alcove in the library, but their method of manufacture seems to be more mysterious than that of any of Henry David Thoreau's literary works.

Several reasonable explanations can be given for the obscurity of Thoreau's involvement with the development of the modern pencil and his work in the family business. One is that, as Petroski pointed out, engineers tend not to be writers and particularly tend not to write about their inventions. This, however, seems a weak explanation in this case. Rather, scholars who write about Thoreau have been interested primarily in his contributions to literature and to ideas about nature; therefore, they have not been concerned with this aspect of his life and, understandably, have omitted mention of it.

But Thoreau's involvement with the pencil and his engineering talent and success regarding one of the most important tools of writers is important in the context of this book: it touches especially on the theme of the connection between civilization and nature and also on the theme of how we learn about nature and make contact with it. Thoreau's work in the family business reveals an important but little-known side of him: his ability to deal with technical observations and to imagine, invent, and improve technologies. This is an unusual talent for an authority on nature. He tackled the problem of purifying graphite as would an engineer—not with a mythology about the mineral but with observations, tinkering, design, and invention.

An engineer's need to understand and to obtain knowledge was echoed in Thoreau's methodology for the study of nature, just as his experience in surveying was echoed in his approach to studying Walden Pond and much else that he discovered in nature. His climb up Mount Katahdin to test a theory about nature was consistent with an engineer's and scientist's approach. His interest in repeated observation, his journal entries, and the tables of data that were to occupy the later part of his life were also consistent with these approaches. Thoreau was a person with a natural talent and inclination for engineering who became captivated with nature; thus, his approach to nature differed from that of, say, someone whose abilities were only in literature and who had an urbanite's arm's-length appreciation of natural surroundings.

Thoreau's wide range of interests and abilities, which allowed him to approach nature as a scientist, surveyor, engineer, and writer, is rare today. Perhaps some of this rarity is due to differences between the early nineteenth century and our time, three of which are relevant here. First, science today is a profession, whereas when Thoreau was alive, it was just emerging as such. Second, modern science encompasses many professions, whereas in Thoreau's time, science was "natural philosophy," not so well separated into a set of disciplines. Third, science today is so well established, authoritative, and well respected that we tend to accept what a scientist tells us at face value, whereas in Thoreau's time, scientists were not necessarily believed to be authorities.

Nowadays, it is difficult to make a contribution to science without professional credentials. A recent article in the *New York Times* noted that "scientific research is so specialized, and its hot fields so crowded, that it is highly unusual for an amateur to make a significant contribution." The article described the work of Gunter Wächtershaüser, a patent lawyer in Germany who had published an article about the origin of life in the well-respected journal *Science*. But even the definition of an amateur scientist has changed: Wächtershaüser, with a Ph.D. degree in organic chemistry from the University of Marburg, was an amateur scientist only in the sense that he did not practice science as a paid professional; he was not an amateur in the sense of lacking appropriate training and knowledge. He "played" at science the way athletes used to qualify for the Olympic Games: with expertise, excellence, and practice, but not having played for pay.

Whereas today it seems a normal thing to seek the opinions and advice of professional scientists about environmental and other societal issues, Thoreau lived in a time when there were few professional scientists. Science was a new but largely amateur activity, in the sense that few made their living as scientists. The transition to modern science was taking place during Thoreau's lifetime: science was becoming a profession. However, Thoreau, who practiced science, never did so as a professional in the sense we use that word today.

Jean Louis Rodolphe Agassiz, born in Switzerland just ten years before Thoreau was born in Massachusetts, came to the United States in 1846, when Thoreau was living at Walden Pond, and subsequently taught at Harvard University as one of the first professional biological and geological scientists in North America. At that time, when science was still called "natural philosophy," people spoke of studying nature rather than studying such narrow disciplines as climate dynamics, atmospheric chemistry, geochemistry, ecology, or chemical and physical oceanography. It was a time when anyone with an interest and an inclination could do science without specific formal training. Anyone could be, or at least claim to be, a scientific expert.

Today, in many scientific fields, such as molecular genetics or high-energy physics, the difference between an expert and a layperson is clear. In these fields, the experts speak a language that is babble to the rest of us but whose results are incredibly powerful, enabling scientists to clone sheep or create bombs that can destroy a city. The distinction between expert and layperson is not so clear, however, when it comes to sciences of the environment—sciences having to do with nature. The terminologies used in environmental sciences, especially in ecology, are more or less familiar and commonplace. This tends to lead people to believe that anyone can be an expert about nature.

Aside from the great difference in terminology, there is a certain similarity between environmental science and the practice of medicine (as distinct from medical research, which is taken as a realm of scientific expertise), insofar as the way the public views both fields. For incurable diseases, such as many forms of cancer, people turn to homeopathic medicine and home remedies when standard medical practice fails. Similarly, with environmental issues, there is often some confusion between obtaining knowledge by "being there and doing things" (analogous to the practice of medicine) and being an expert in the search for new knowledge (analogous to doing medical research).

In both these areas of practice—medicine and environmental science—people "play" as amateurs in the classical sense, as I have used the term. There are many naturalists with great expertise in identification of species and in the structure and form of nature. As a result, in public forums it is often unclear who is and who is not an expert. In this way, our views of environmental issues have not progressed beyond Thoreau's time. The knowledge has changed, but the issues are timeless.

The first half of the nineteenth century was also a very different time from ours in terms of technology. The age of steel and steam was just beginning; the era of rubber-coated copper wires, lead-acid batteries, and electric generators was far in the future. The bicycle was a new thing in Thoreau's time. The first bicycle to use pedals, cranks, drive rods, and handlebars was introduced in Scotland in 1839, and the kind of bicycle we take for granted, with a chain drive and front and rear wheels of equal size, was not invented until 1885, more than twenty years after Thoreau's death. The pneumatic tire came along even later, in 1888, invented by John B. Dunlop. True, the steam engine had been around for a while—the patent on James Watt's steam engine was already seven decades old at the time Thoreau sat in his cabin at Walden Pond. But the first railway to operate in the United States, the Baltimore and Ohio Railroad, ran its first trains when Thoreau was eleven years old, in 1828, and the golden spike that completed the transcontinental railroad of the Union Pacific was not hammered into place until 1869, when Thoreau had been in his grave for seven years. As a result, the perception of technology and its power in Thoreau's time was different from—much humbler than—our perception today.

Since Thoreau's time, technology has enabled people to exert greater and greater control over local environments. We live with an increasing sense of independence, even isolation, from nature and its vicissitudes as a result of the confidence instilled in us by devices such as central heating and air conditioning systems. Lacking these means of local environmental control, people in Thoreau's time probably felt closer to nature on a day-to-day basis than do the majority of urban and suburban residents who make up most of the world's human population today.

Our modern society gives us a sense of technological security, the sense that with our continually advancing technology we are increasingly more protected from nature's extremes. But it is an ironic and flawed sense of security. The fallacy was perhaps best pointed out by Lewis Mumford, a great historian of cities. Mumford wrote that although the artificiality of cities gives their citizens a sense of independence from the surrounding environment, that very artificiality makes cities ever more dependent on their surrounding natural resources and ever more sensitive to environmental change. So it is with us, cocooned within our double-glazed windows and 6× insulation. This sense of isolation from nature reinforces the idea, prevalent today, that nature is "out there" and that preserving nature is generally an activity that takes place over the horizon or at those locations one seeks to visit on vacation but with which one never becomes familiar or intimate. We believe that we go to nature, but we are not part of it. We no longer see ourselves as immersed in nature, even though we are.

As a result, the conservation and protection of wild living resources is typically seen as an activity that is beyond our day-to-day urban experience, but it is one that we nonetheless take on ourselves, with or without direct contact with that nature.

Today, there is a generally accepted hierarchy of prestige regarding creativity in science and engineering. Fundamental scientific research is given the highest prestige. It is generally believed in our society that the greatest imaginative leaps, with the greatest importance to civilization, come from this kind of science, which according to our mythology operates totally apart from practical questions and problems. Next in the hierarchy is applied science, which begins with a practical question, such as how to cure cancer, and seeks a scientific answer. Lowest in the hierarchy is engineering, which is viewed as a technical exercise carried out without great leaps of imagination or insight. According to this hierarchy, inventions of practical value come from a series of advances that begin with fundamental science, move to applied science, and eventually reach engineers, who then apply the new ideas. The reality is more complex. A strong case can be made that the motivation for much significant fundamental science comes from attempts to solve practical problems. Much of the development of thermodynamics came about in the nineteenth century from an interest in developing a more efficient engine. The research that arose from this practical question led to development of the rich and important fundamental science of thermodynamics, which in turn gave birth to the diesel engine.

My own experience in environmental sciences suggests that this standard hierarchy is incorrect. Environmental sciences have progressed much more in accordance with the model of the development of thermodynamics. The study of nature is so complex and difficult that without practical questions to answer, the field tends to wander into obscure corners that become dead ends. Much that is productive in environmental science comes from attempts to solve real-world problems. For example, the motivation for ecosystem research, in the sense of large-scale funding and recognition of its importance, originally came largely from post–World War II attempts to understand potential environmental effects of nuclear war. Concern with birds becoming endangered from the use of DDT led to an understanding of the concentration of oil-soluble compounds in food chains. Later, significant funding for ecosystem science came from a need to understand the effects of clear-cutting on forests. Current interest in restoration ecology—restoration of previously damaged and altered ecosystems—is leading to new fundamental insights.

The development of the modern jet-propelled airliner is an interesting example of the progress of technology and the role of innovation and creativity. According to William H. Cook, one of the engineers at the Boeing Company who helped develop the first successful passenger jetliner, the Boeing 707, the development of airplanes from Wilbur and Orville Wright's craft to the Boeing 707 took place with "meager theories" that "were often erroneous." The desire and need to develop aircraft outpaced the development of the theory of fluid dynamics and aeronautics in general. "Intuition played a major role in the creative process," Cook wrote. "As new ideas were developed and tested, new theories were developed to explain the successes and provide a key for optimizing performance even further." Moreover, the individuals who carried out key experiments were viewed as eccentric, and "at first they received no help from the academic or political arenas." Academic institutions, supposedly Western civilization's center of creativity, were not, he suggests, the source of inspiration for the key developments in the invention of the modern airplane.

As with the airplane, technological progress has often involved direct creative insights arising from a struggle to solve a specific practical problem that was in advance of existing theory. Often, such leaps of insight come from people outside accepted societal entities. Again, the development of the modern jetliner is revealing. The jet engine was invented independently by two graduate students, Frank Whittle in England in 1929 and Hans von Ohain in Germany in 1934. The idea of the jet engine had been proposed earlier but rejected as impractical because it seemed too inefficient in the use of fuel. Whittle, a pilot, developed a fuel-efficient jet engine whose key element was a single shaft containing an air compressor and a turbine driven by burning fuel. He explained his design to a friend, also a pilot, who had the financial resources to fund initial development. Thus, the jet engine, the centerpiece of modern aircraft, was the product of "amateur" enthusiasts, initially developed outside the field of aeronautical engineering. The conventional wisdom of the field at the time was that a practical jet engine could never be built; the route to more powerful aircraft was believed to be development of ever more complex piston engines.

By the time of the development of the Boeing 707, the first truly successful modern passenger jetliner, the military had considerable experience in the use of jet engines. However, few engineers believed that the jet engine could be used successfully for passenger travel. Cook describes a process of tinkering and imaginative thinking, as well as some luck and fortune, that solved several problems as the 707 was developed. These problems were beyond the reach of contemporary theory about flight approaching the speed of sound; therefore, engineers had to rely on experimentation, experience, and insight.

Thoreau excelled at this kind of process, applying it to his improvements of the pencil as well as his study of nature. It is an approach that is difficult to pursue when the field of study is heavily laden with deep-seated myths and the required insights run counter to the prevailing paradigms of academic institutions. Our understanding of nature and its connection to human beings and civilization has suffered from these heavy pressures.

Innovation in technology and in solving environmental problems often seems to come from iconoclastic individuals working outside of established institutions and established paradigms. Thoreau seemed to understand this and pursued a different path, often working outside the establishment and never becoming a professional or a long-term part of established institutions. This is a lonely approach. It is this quality, more than a lack of sociability, that makes Thoreau appear to have been a loner, operating outside of society. Yet Thoreau's life and his approach to nature and technology reveal something about the requirements for creativity and innovation. By their very nature, innovation and creativity—the discovery of something new—must oppose some established ideas and beliefs. Perhaps we will always need a Thoreau to help us progress beyond standard paradigms.

Thoreau's combination of experience and knowledge seems all the more unusual today because of the common prejudice that technology is a force in opposition to nature; it is difficult to conceive of appreciating both nature and technology simultaneously. Thoreau is not alone, however, among writers. The lyrical and beautiful nature writing in Antoine de Saint-Exupéry's *Wind, Sand, and Stars,* for example, shares this combination and discusses it explicitly. One of the first French pilots, Saint-Exupéry flew some of the first airmail planes between France and North Africa. In *Wind, Sand, and Stars,* he wrote about flying alone at night in a small single-engine airplane. He felt immersed in nature in a wonderful and profound way, both because he, as captain of his fragile machine, had to understand nature and its forces or die and because the night flights gave him a beautiful view of the Earth. From these flights, looking out from the cockpit beyond the red glow of his instrument panel, beyond the drone of the lone engine that was all that lay between him and disaster, he created a great work expressing the beauty of the Earth. The machine, for Saint-Exupéry, was the path to understanding and becoming part of nature in a new and beautiful way.

Joseph Conrad also wrote about the effects of technology on a person's perception of his relationship with nature. In his short story "Typhoon," Conrad describes the ocean journey of a newly invented steamship whose captain believed that the power of this new technology empowered him to ignore even the most severe storms. In his drive for efficiency, he took the ship on a direct path through a hurricane, which almost destroyed it.

Conrad describes the storm primarily from the point of view of the passengers in the depths of the ship, who cannot see out but are thrown about helplessly. Physical nature's power is expressed through its effects on individuals. Conrad's story reveals his preference for the age of sail, but it also shows how a falsely placed confidence in technology, an arrogance generated by that technology, can lead a human being astray.

Our cultural, scientific, and technological background defines to a large extent, and therefore restricts, our perception of nature and ourselves. In our time, we tend to view environmental concerns as if they were entirely new to civilization. In popular writings, the pressures of human population growth and modern technology appear to have created entirely new issues. But anyone who has read the classics, as had Thoreau, sees things quite differently. In his classic book *Traces on the Rhodian Shore,* Clarence J. Glacken, a great geographer and historian of the idea of nature, discusses the continuing concern of writers, philosophers, theologians, and early scientists with the relationship between people and nature, a concern that dates back several thousand years.

I read once about a ship that was built in medieval times with masts elaborately decorated with carved gold. Compare that with the sleek "form follows function" design of modern sailing vessels, especially those built for races. The medieval ship was a product of a time when people were emerging from dark dependence on a hostile world over which they had little control, the world of *Beowulf,* into a world in which technology was permitting a slightly more secure existence. Therefore, one would gild the mast of his ship to prove his wealth and power over nature. In contrast, today a person who wishes to win the America's Cup race demonstrates his wealth and prowess by commissioning the most functionally designed vessel and sailing it with the greatest skill. These two kinds of vessels, in two different ages, symbolize a difference in the perception of sense of the connectedness between people and nature. They also reinforce the persistence of the importance of that relationship.

Humanity's concern with its relationship with nature is universal, but whether an individual perceives himself at the mercy of nature and "under its thumb," sees himself at peace with and a part of nature, or believes himself to be above nature and in control of it, to the point of destroying nature, depends on the power of available technology. The great technological power of the present age has placed us in a unique position in relation to nature. Thoreau was perhaps able to feel at home within nature because of his competence in and familiarity with the technology of his day. I suggest that the path to compatibility between civilization and nature requires that we similarly see ourselves within nature, neither under its thumb nor nature under ours. This is one of the lessons of Thoreau's involvement in developing the pencil.

The development of the pencil and the development of the jetliner were design competitions, informal and formal, rather than searches for a single truth. Although in retrospect the successful design may seem inevitable and the only possible one, that is never the way things look to those involved in the struggle to invent and solve. Today, we are confronted with a new set of problems. Biological nature is, of course, a set of systems of greater complexity than either the pencil or the jet aircraft; thus, our ability to solve environmental problems and find a sustaining connection between nature and civilization requires an even greater inventiveness—one that is hard to obtain under a burden of established but incorrect paradigms and of institutions with vested interests in outmoded approaches. For those of us today who consider ourselves environmentalists or lovers of nature, one key to Thoreau's relevance is his ability to break through the prevailing views of his time, perhaps because he had the natural inclination of an engineer and tinkerer. It is something for us to ponder as we struggle to find ways to sustain both nature and civilization.

Chapter 9

Breakfasting on Cape Cod

Som'body better help me, 'cause I can't help myself.
Junior Wells

"The Cape is merely a temporary deposit of glacial sediment, pausing briefly in the geological sense on its way to inevitably being washed into the sea."
Graham Giese, oceanographer,
Center for Coastal Studies, Provincetown, Massachusetts

ON OCTOBER 10, 1849, Henry David Thoreau arrived at Orleans, Massachusetts, a small village about halfway out on the peninsula of Cape Cod. He had traveled by train from Boston to Sandwich, the village nearest the mainland end of the Cape, and then by stagecoach to the middle of the Cape. From there, he began walking to Provincetown, with poet Ellery Channing as his companion. Provincetown, at the tip of the peninsula, would have been about twenty-five miles away by a direct path. But they took a meandering route. They went east to the shore, about a five-mile walk from where they left the stagecoach, and came to Nauset Beach, just east of the village of Orleans. There, Thoreau got his first view of the ocean.

Today, Nauset Beach is part of the Cape Cod National Seashore and on a sunny day affords a pleasant view of the ocean side of the Cape. From a wooden walkway near the visitors' center, one can catch the salt breezes that rise over the land. Below are gentle dunes covered with dune grass and a scattering of coastal shrubs. Sandbars stretch for a long distance; gulls call over the sea and sand. Water and land, at nearly the same elevation, compete for dominance.

This was the first of four trips Thoreau made to the Cape. He made a second trip the next year, in June 1850, by himself; a third in July 1855, again with Channing; and a final trip by himself in June 1857. He integrated all these trips into a single narrative in his book *Cape Cod.* These visits had a great effect on Thoreau, making him aware that the Cape was a shifting, dynamic landscape; allowing him to experience a new kind of wildness, the ocean; and leading him to think about nature in new ways. But he did not make these discoveries totally on his own. He was introduced to new ideas by local experts.

Thoreau developed a specific approach to the use of local expertise. He pondered the question often during his life: Whom should one trust regarding knowledge of nature? Whom should one accept as an expert? In general, he preferred to observe and learn for himself, to trust himself, but he also recognized that this was often not possible and sometimes not sufficient.

In the Maine woods, Joe Polis, his guide on his last trip there, was the expert. When Thoreau and Ed Hoar were sent out on their own, their understanding was not sufficient; they needed their expert. Had Polis been with them as they struggled through Umbazooksus Swamp, he would have found a way through; his expertise would have worked. But Polis's directions to Thoreau had not been helpful; the expert's advice alone was not sufficient.

In our era of modern science and technology, it is often unclear who knows the truth about the highly technical issues involving nature, such as the causes of environmental pollution or the likelihood of extinction of an endangered species. Even so, we seek experts. On the one hand, we tend to trust science so much that we ask scientists to tell us what to do—to make policy. On the other hand, we tend to distrust science as a basis for policy. We reject science in part because it has led to the very technology that allows us to destroy the environment. Science is viewed as the root of the problem. Thus, we simultaneously admire scientists and are skeptical of them—we believe that science is objective but are concerned that scientists may be "hired hands" who will manipulate data to provide whatever answer their funders want. Society is struggling to understand the role of science and scientists in helping us solve environmental problems and increase our understanding of nature. We should remember that science has not been free from ethical and moral assumptions about nature; indeed, the science of ecology—our science of nature-knowledge—has been greatly influenced by belief systems. As environmental issues heighten in intensity, our trust or distrust of scientists as experts becomes more and more important.

Viewed on a map, Cape Cod is shaped like an upraised left arm of Massachusetts, fairly straight to the elbow and then curving upward, the hand pointed west, back toward the mainland shore. The distance from the modern Cape Cod Canal—a canal just northwest of Sandwich that marks the present beginning of the Cape—to Provincetown at the tip is approximately sixty miles by highway. In this way, the Cape forms a bay to the west, which is hydrologically part of Boston Harbor; the eastern side of the Cape is open to the Atlantic Ocean.

After walking about eight miles along the ocean side of the beach, Thoreau and Channing turned inland to the wrist of the Cape and came upon some houses "uncommonly near the eastern coast" between the towns of Wellfleet and Truro. Thoreau described the land near the ocean as patches of bayberry "straggled into the sand" and shrub oak. In those days, the Cape was not thought of as a destination for vacation or recreation. There were few formal inns or other places to stay; one either camped or sought to spend a night in the home of a resident. Thoreau knocked on the door of one of the few houses, but nobody was home. Then he knocked on a second door and "a grizzly-looking man appeared." After asking a few questions, the man invited Thoreau and his traveling companion to come inside.

The old man said his name was John Young Newcomb. He told Thoreau that he was eighty-eight years of age, old enough to have heard the guns fired at the Battle of Bunker Hill. In his journal, Thoreau referred to Newcomb as the "Wellfleet Oysterman" and described him as "the merriest old man that we had ever seen, and one of the best preserved." Newcomb's house still stands today, a rambling white clapboard house with old apple trees around it. The travelers talked with Newcomb for a long time about many things, including the Oysterman's special apples and the history of settlement on the Cape.

In the morning, the Oysterman's wife cooked breakfast over the fire-

place. The Oysterman told stories while breakfast was being prepared, chewing tobacco all the while "and ejecting his tobacco juice right and left into the fire behind him," wrote Thoreau, "without regard to the various dishes which were there preparing." Watching the destinations of the tobacco juice, Thoreau chose to eat applesauce and doughnuts, which he "thought had sustained the least detriment from the old man's shots." His traveling companion ate hotcakes and green beans, "which had appeared to him to occupy the safest part of the hearth," Thoreau continued. "But on comparing notes afterward, I told him that the buttermilk cake was particularly exposed, and I saw how it suffered repeatedly." His companion responded that the applesauce had been "repeatedly injured." The Oysterman, grizzled and spitting tobacco juice on his guests' breakfast foods, had few of the qualities one might attribute to an expert, but in spite of the old man's appearance and habits, Thoreau listened carefully to what he had to say:

> The old Oysterman had told us that many years ago he lost a "crittur" by her being mired in a swamp near the Atlantic side east of his house, and twenty years ago he lost the swamp itself entirely, but has since seen signs of it appearing on the beach. He also said that he had seen cedar stumps "as big as cart-wheels" (!) On the bottom of the bay, three miles

> off Billingsgate Point, when leaning over the side of his boat in pleasant weather, and that that was dry land not long ago.

Trees had once grown where now was ocean. According to the Oysterman, the Cape was not stationary; the very land on the Cape moved.

This idea intrigued Thoreau. He wrote that another person "told us that a log canoe known to have been buried many years before on the Bay side at East harbor in Truro, where the Cape is extremely narrow, appeared at length on the Atlantic side, the Cape having rolled over it, and an old woman said—'Now, you see, it is true what I told you, that the cape is moving.'" The idea that the land surface of the Earth was dynamic was novel at that time. More prevalent was the belief that life and its local environment—oaks, bayberries, the soil, bedrock—were fixed and created a permanent, static setting.

Thoreau did not simply accept as truth the stories of the Wellfleet Oysterman and the Old Woman, as he called her. As he continued to walk east and north on the Cape, he reached the Highland Light, where he talked with the lighthouse keeper. The Highland Light was a landmark then as it is today, a classical white pillar rising above a white building on the edge of a picturesque dune high above the beach and the water, facing east toward the open Atlantic Ocean. It sits on the edge of an undulating

landscape of dune grass, shrubs, small oaks, and pitch pines, a mixture of patches of grassland, shrublands, and open woodlands stunted by salt spray. It is a lonely but picturesque landscape. The huge dune supporting the lighthouse affords a grand view of the shore below. From the lighthouse, the dune falls away steeply for a long distance; far below, people strolling along the strand resemble tiny toy figures.

The lighthouse was built in 1798 to guide ships away from dangerous shoals along the coast of the Cape, and it performed that function during Thoreau's time. Today, the lighthouse is automated and no longer has a keeper.

To Thoreau, the lighthouse keeper was a different sort of expert from the Wellfleet Oysterman and the Old Woman. These three people typify two kinds of experts: professional and experiential. A professional expert has obtained specialized training or education in a subject and, in a scientific age, makes use of quantitative measurements. An experiential expert has experienced something firsthand and therefore has qualitative but direct and often emotional knowledge about it, with or without training or education. A meteorologist, for example, can be a professional expert on tornadoes; a person who has been in the path of a tornado and has suffered its effects and survived is an experiential expert on tornadoes. A meteorologist studying tornadoes in the field who becomes caught up in one could be both kinds of expert. The Oysterman and the Old Woman were experiential experts. They lived on the Cape and knew it from observing it as they went about their lives. The lighthouse keeper was a professional expert in that his job required him to know about the Cape and he had done so for a long time. He had lived there for sixty years, a length of time that amazed Thoreau.

Like the Wellfleet Oysterman and the Old Woman, the lighthouse keeper had observed the erosion and movement of the Cape. "According to the light-house keeper, the Cape is wasting here on both sides, though most on the eastern," Thoreau wrote. Because the eastern shore was the ocean side of the Cape at this location, the lighthouse keeper's comment confirmed what Thoreau had heard from the Oysterman and the Old Woman.

But Thoreau did not leave things at that. He did not accept the opinion of either kind of expert, professional or experiential, without conducting tests of his own. At the time, the Highland Light stood back from the edge of the dune a distance that Thoreau wrote was 330 feet. As with the depth of Walden Pond, he made measurements. Having worked as a surveyor, Thoreau improvised surveying equipment in order to do so. "I borrowed the plane and square, level and dividers, of a carpenter who was shingling a barn near by," he wrote, "and using one of those shingles made of a mast, contrived a rude sort of quadrant, with pins for sights and pivots, and got the angle of elevation of the Bank opposite the light-house, and with a couple of cod-lines the length of its slope, and so measured its height on the shingle." He observed that the dune rose 110 feet "above its immediate base" and 123 feet above mean low tide.

Next, he checked his measurements against those of other land surveyors. "Graham, who has carefully surveyed the extremity of the Cape, makes it one hundred and thirty feet," he wrote.

Then he looked for signs of erosion—making qualitative observations. He found evidence of erosion about one-half mile south of the lighthouse, at the point of highest land in the vicinity. There, he saw streams "trickling down [the dune] at intervals of two or three rods," which left erosional shapes like "steep Gothic roofs fifty feet high or more." At one location, the shapes were "curiously eaten out in the form of a large semicircular crater."

Still not content with either the lighthouse keeper's opinion or his own measurements, he examined data kept by the lighthouse keeper. "We calculated, *from his data,* how soon the Cape would be quite worn away," Thoreau wrote.

Thoreau made additional measurements when he returned to the Cape the following summer. "Between this October and June of the next year," he wrote, "I found that the bank had lost about forty feet in one place, opposite the light-house." From these observations, he concluded that Cape Cod was wearing away at a rate of about six feet per year. But he was cautious about simply extrapolating and generalizing from a few observations. "Any conclusion drawn from the observations of a few years or one generation only are likely to prove false," he wrote, "and the Cape may balk expectation by its durability." Such skepticism—even about one's own measurements and observations—is one of the important features of science and of scientists.

From the observations of local experts and the results of his own investigations, Thoreau began to generalize about the dynamics of the geology of Cape Cod. "On the eastern side the sea appears to be everywhere encroaching on the land," he wrote. "Not only the land is undermined, and its ruins carried off by the currents, but the sand is blown from the beach directly up the steep bank."

Thoreau began to understand that nature is dynamic. This was an important moment in the development of ecology and modern environmentalism because the idea of the naturalness of change ran counter to the great, ancient myth of the balance of nature, which, before and during Thoreau's time, was the accepted explanation of how nature worked. Today, it remains a powerful idea and continues to be the foundation of much environmental law, policy, and beliefs. Central to this myth is the belief that nature, if left alone, achieves a constancy of form and structure, and if disturbed, it will eventually return to that same form and structure. But if this myth were true, Cape Cod could not have moved and would not have rolled over a sunken log canoe.

"Erelong, the light-house must be moved," Thoreau wrote, because the hill dunes would erode back far enough that the lighthouse itself could fall down the slope. This prediction came true, as any visitor to the Highland Light today can tell you. In the mid-1990s the lighthouse was fenced off because it was deemed unsafe to visit: it was now so near the edge of the dune that it was in danger of tumbling over. Now it has been moved back from the edge. For a while the historic and beautiful landmark could be viewed only from outside an ugly chain-link fence. In contrast to the remoteness of the lighthouse in Thoreau's day, it is now reached easily by road. Tour buses and private cars park just inland so that tourists can view the building.

We know today that Cape Cod is a product of the continental glaciers. In Thoreau's time, Jean Louis Rodolphe Agassiz and other scientists had only recently discovered that continental-scale glaciations had subjected the Earth's surface to major alterations. The Cape is in part the moraine of the last North American continental glacier—stones and earth of all sizes, from boulders to tiny particles of clay, that the glacier carried as it moved forward, ultimately depositing the material at its southernmost limit. The moraine formed the inner, bay side of the Cape; thus, the soil on this side, which faces the mainland, toward Boston and Concord, is richer than that on the outer, ocean side.

The rest of the Cape is glacial outwash, material deposited by rapidly moving waters that flow from a glacier as it melts and retreats. The flowing water segregates by size the materials it carries: heavier, larger materials are carried only by the most rapidly moving water, so these are deposited first as the water loses energy and speed. The first materials to be deposited are rocks and pebbles; next are sands, then silts. Finally, where the waters become almost motionless, clays are left. Outwash soil is typically sandy and usually infertile. That, along with the presence of salt spray from the ocean, which kills the growing tips of plants, is one reason why the vegetation of Cape Cod is generally low, scrubby woodland, shrubland, and grassland.

Both the moraine and the outwash are easily eroded and redeposited by the ocean. That is why the Cape is shaped like an upraised arm. When first formed, at the end of the last ice age, the Cape was more or less a straight peninsula tending east to west. But ocean currents and winds act more forcefully the farther out to sea the Cape extends, and over the thousands of years since the ice melted back, the Cape has moved ever east and north at its outer part, forming the "elbow." It has rolled over itself for thousands of years.

After Thoreau visited the Highland Light and made measurements there, he saw other evidence of how the Cape had shifted and moved. "The entrance to Nauset harbor, which was once in Eastham, has now traveled south into Orleans," he wrote. "The islands in Wellfleet harbor once formed a continuous beach, though now small vessels pass between them," so that "perhaps what the Ocean takes from one part of the Cape it gives to another."

The ocean washes sand away, and the wind blows it back. "If you sit on the edge you will have ocular demonstration of this by soon getting your eyes full," he wrote. "The sand is steadily traveling westward at a rapid rate," he continued, "so that in some places peat-meadows are buried deep under the sand, and the peat is cut through it; and in one place a large peat-meadow has made its appearance on the shore in the bank covered many feet deep, and peat has been cut there." He described the effects of storms that move large portions of sand. A few months before one of his previous trips, one such storm had "cast up a bar" near the Highland Light that was "two miles long and ten rods wide," leaving a narrow cove that was good for swimming. But during a later trip, Thoreau found that the bar had moved north. "The sea thus plays with the land holding a sand-bar in its mouth awhile before it swallows it, as a cat plays with a mouse; but the fatal grip is sure to come at last."

Later in life, Thoreau pursued the idea of a dynamic nature elsewhere. In a report to the state of Massachusetts, he was one of the first to use the term *succession* for vegetation in the sense ecologists use it today—to mean the process by which a forest revegetates and reoccupies a location that has been cleared by fire, storms, or human action.

Thoreau not only learned on Cape Cod that the ground moved as a result of the physical forces of wind and water; he also recognized that life tended to oppose those forces, stabilizing the soil, at least for a brief time. He saw that dune grass tended to stabilize the dunes. He quoted from a "Description of the Eastern Coast," which said that "beach grass" grew "about two feet and a half" during spring and summer, and then

> the storms of autumn and winter heap up the sand on all sides, and cause it to rise nearly to the top of the plant. In the ensuing spring the grass mounts anew; is again covered with sand in the winter; and thus a hill or ridge continues to ascend as long as there is a sufficient base to support it, or till the circumscribing sand, being also covered with beach-grass, will no longer yield to the force of the winds.

He also noted that Timothy Dwight had written in his *Travels in New England and New York* that "inhabitants of Truro were formerly regularly warned under the authority of law in the month of April yearly, to plant beach-grass, as elsewhere they are warned to repair the highways." The residents had built a road on a narrow part of the Cape between Truro and Provincetown, "where the sea broke over in the last century." The road was made by planting dune grass. "Thus Cape cod is anchored to the heavens, as it were, by a myriad little cables of beach-grass, and, if they should fail, would become a total wreck, and erelong go to the bottom."

The dynamics of nature—the tendency for nature's physical forces to erode while its biological forces tend to stabilize—was a novel idea in Thoreau's time. The role of dune grass became better known to ecological science about half a century after Thoreau's death, when the dunes along the shores of the Great Lakes were studied. There also, dune grass stabilized the sand while wind and water eroded and moved sediments. The constant battle on the dunes between life, which tends to build up, and the physical forces of the environment, which tend to destroy, led to some of the earliest recognition of the dynamics of nature from the new science of ecology that emerged at the beginning of the twentieth century.

Dune grass sends out runners that form a latticework—a living infrastructure—to hold the sand in place. Once the sand is stable, seeds of other plants have time to germinate and grow without being blown away. In time, a forest develops. But this forest is not permanent. Eventually, wind and waves do their damage and the layers shift. The dynamics of the dunes on Cape Cod are very clear to us because they happen within the span of a human life and at a scale we can recognize. It is this layering of changes, one on another, that helps us begin to understand the great complexity of natural systems. Today, we are building on that idea as we undergo a revolution in our understanding of the environment. We now see that there are various kinds of changes, one layered on top of another, even on a global scale.

Modern Cape Cod is much changed, of course, from when Thoreau visited it. When he was there, the Cape was not considered to be the beautiful recreational destination it is today.

In the 1970s, I spent three years on Cape Cod, living in Falmouth and working in nearby Woods Hole, at the Marine Biological Laboratory at the Cape's southwestern extreme. Woods Hole is a point of departure for the ferries that take tourists to Martha's Vineyard and Nantucket and the home of several major marine research centers. The Marine Biological Laboratory is the oldest of these, having been founded in the latter part of the nineteenth century, within a few decades of Thoreau's death. There is also the Woods Hole Oceanographic Institution, a center for deep-sea research as well as the location of a U.S. Geological Survey office and a National Marine Fisheries Service research center. Falmouth is the nearest large residential and commercial center, with a main street along which tourist traffic crawls in the summer. A drive down Falmouth's main street leaves the impression that our civilization has created something permanent there—macadam roads, a school and playing field, stores of cement, a bowling alley. But on the Cape, this apparent permanency is only an illusion. Like the log canoe rolled over by the Cape, these structures will eventually be subjected to the irresistible forces of wind and water.

Thoreau became well aware of the temporary nature of our technological structures on the Cape and the battle between the forces of wind and water and the constructive activities of people. "Sand is the great enemy here," he wrote. "The sands drift like snow. There was a school house, just under the hill, filled with sand up to the tops of the desks, and of course the master and scholars had fled. Perhaps," he commented wryly, "they had imprudently left the windows open one day, or neglected to mend a broken pane."

Although Thoreau chose to learn directly from experience whenever he could, there were, of course, times when he was not able to do so. This is a problem we all face, and in our time we face it with greater frequency and importance because of the tremendous growth of science and technology. We cannot know all the answers ourselves. We lack the time, interest, and abilities. The question is, If we cannot learn about everything ourselves, whom can we trust to give us answers, directions, advice?

There are three possible groups of individuals to whom we can turn. The first group is contemporary professional experts—people with specific credentials, such as those with a doctoral degree in a science, registered engineers or architects, or physicians—whose credentials suggest that years of education and training qualify them to address problems in a specific technical subject. The second group is the great thinkers of the past, whose writings have stood the test of time. The third is experiential experts—local people with local knowledge based on their experience of living in an area and observing it carefully, such as Thoreau's Wellfleet Oysterman and Old Woman on Cape Cod.

Each kind of expert presents us with certain benefits but also with problems, doubts, hesitancies. We face these problems daily, problems exacerbated by rapid increases in knowledge, division of disciplines, and advances in technology. Some would say that our society is reaching a crisis in its regard for experts. We are caught up in two kinds of problems. The first is, Who really is an expert? The second is, What should we do when experts disagree?

On the one hand, some may accept as an expert anybody who appears on television, hosts a radio talk show, or writes an article that is published in a newspaper. On the other, we grow increasingly skeptical of scientists, engineers, physicians, and others with real technical expertise who appear in court, testify before Congress, or are quoted in the media.

An article by R. B. Schmitt published recently in the *Wall Street Journal*, titled "Witness Stand: Who Is an Expert? In Some Courtrooms, the Answer Is 'Nobody,'" illustrates the problem. "Experts have been the brains behind the growth in injury lawsuits since the 1960s," Schmitt wrote. "Over the years, they have become ubiquitous in court, indispensable in many suits, and a flourishing industry in their own right. For a fee, experts will do most anything, or so it seems, from identifying the design flaws in gas-barbecue grills to diagnosing toxic side effects from industrial accidents." In the past, Schmitt continued, judges left it to the juries to determine whether an expert was credible. He quoted Joseph Sanders, a law professor at the University of Houston: "The view from the bench has been, 'What the hell! He has a Ph.D. from MIT.'" But today, that is changing; across the country, judges are throwing out experts as unqualified. Alan Milner, a metallurgist who since the 1970s had testified in some 350 court cases about defective automobile tires, was banned from a courtroom in St. Louis in the fall of 1996 when the judge decided he was not qualified to testify in a case involving a tire that blew up in the face of a mechanic. Confronted with experts representing both sides in cases, judges are beginning to wonder who really is an expert.

Television confuses us in another way about who is an expert. During recent floods on the Mississippi, Missouri, and Red Rivers, eyewitness news accounts brought us the immediacy of interviews with local people whose homes were flooded, farmland inundated, and crops destroyed. They were local, experiential experts—they knew what it was like to be victims of the disaster. But the interviewers' questions slid from what it was like to be there to what were the causes of the floods and what should be done to prevent them. These last two questions require a different kind of expertise—scientific and technical. But the interviewers confused the two kinds of experts.

Such confusion raises a major problem in issues related to nature and the environment because all of us are experiential experts about what it is like to be within some kind of environment. There is an analogy with medical care. Each of us alone knows what it is like to be within our bodies and experience our own well-being or pain. We are experiential experts about our bodies, but we are not technical experts in the medical treatment of our bodies. Yet without even consulting a physician, we can obtain a wide spectrum of health remedies, from home remedies handed down from generation to generation in a family to homeopathic medicines available over the counter to pharmaceuticals that are the result of extensive scientific research and testing. People may try all these kinds of remedies for incurable or difficult-to-cure diseases such as cancer, and popular ones come in and out of fashion.

Or, suppose you are in search of the optimal diet. You have a friend who has lost weight and kept his blood cholesterol level down by sticking to a high-protein diet; is he the expert to consult? Or would it be the physician who wrote the latest fashionable diet best-seller, on display at the airport candy store counter? Or perhaps your personal physician, who may or may not have been well trained in nutrition? Should you place your trust in the handsome actor wearing a white coat on television, hawking vitamins? Or should you follow a commercial weight-loss program? The answer to the question of who is the dependable expert is not easy to determine. Thoreau faced a similar dilemma over whom to trust and when to trust somebody else's opinion. Again, his experiences and approach suggest a process that can help us today.

Thoreau used local expertise as a starting point for his own thinking, obtaining insights that he would then subject to further testing. He was selective about whom he listened to—on the question of land movement on Cape Cod, it was longtime residents, not just anybody who happened to be standing by the edge of the dune. The two kinds of experts, experiential and professional, suggested possibilities that formed a starting point for his own investigations. Then he sought quantities: quantitative measurements he could make for himself as well as the measurements the lighthouse keeper had made. The fact that the lighthouse keeper had made measurements reinforced his status as a professional expert, in contrast with the Oysterman and the Old Woman. Next, convinced about the observations, Thoreau thought about their general implications. Just as he had done in measuring the depth of Walden Pond, Thoreau went beyond the limits of a single set of observations and sought a more general theory.

As Thoreau's experiences illustrate, there is value in the knowledge of local people who may lack formal training in a specific discipline. But there is also a problem. You can always find somebody in an area who will express almost any opinion you seek. How do you differentiate those who know what they are talking about from those who do not? Thoreau is clear about this: use the opinions of local experts as a starting point for your own exploration, and seek out those who have made quantitative observations that go beyond opinion and casual observation. In describing Thoreau's accounts of the Highland Light and the movement of Cape Cod, I am using him as a professional expert on the history of the Cape, in the same way he used the lighthouse keeper. Thoreau was there and knew how to survey. He provided the quantitative results of his measurements and presented his ideas for us to evaluate. The approach is clear. First, regardless of the fashionableness of your experiential experts, if they appear to have extensive direct experience, listen to what they say. Second, test their ideas yourself, using quantitative methods, or, third, if you cannot do the tests yourself, seek a professional expert who can.

At Walden Pond, Thoreau was both the experiential expert and the professional expert. On Cape Cod, he was simply a tourist who listened carefully to the local, experiential experts, those who had lived there for a long time and thought about the Cape as an environment. He then picked up the investigation with the process he had used at Walden Pond—the scientific method.

However, he went an additional step beyond the onetime, static measurements of depth he made at various points in Walden Pond, recognizing that measurements made at a single point in time might not be representative of a dynamic environment such as the dunes. At the pond, the questions had to do with shape and form that changed slowly in relation to his lifetime; on the Cape, he was confronted with the dynamic character of nature, which we today are only beginning to accept in our approach to managing and conserving nature. Confronted with the Cape's rapid changes in shape brought about by wind, waves, and tides, Thoreau realized the need for a series of measurements over time—a step we must add as a part of Thoreau's approach to knowledge of nature.

Thoreau listened to both kinds of experts, experiential and professional, but he listened in a different way to each kind. From the local, experiential expert, he sought informal insights. From the professional expert, he sought insights and formal, sometimes quantitative, knowledge. Thus, Thoreau made use of experts in a specific way: as a source of insights and ideas but not of complete understanding. They were for him a beginning, a source of hypotheses that could be tested using the scientific method—a beginning of new kinds of contact with nature.

Chapter 10

The Sound of a Woodchopper's Ax

"The individual trees of those woods grow up, have their youth, their old age, and a period to their life, and die as we men do. You will see many a Sapling growing up, many an old Tree tottering to its Fall, and many fallen and rotting away, while they are succeeded by others of their kind, just as the Race of Man is. By this Succession of Vegetation the Wilderness is kept clothed with Woods just as the human Species keeps the Earth peopled by its continuing Succession of Generations."

Thomas Pownall, *A Topographical Description of the Dominions of the United States of America*

"What though the woods be cut down. It appears that this emergency was long ago anticipated and provided for by Nature, and the interregnum is not allowed to be a barren one. She not only begins instantly to heal the scar, but she compensates us for the loss and refreshes us with fruits and such as the forest did not produce."

Henry David Thoreau, "Huckleberries"

On the evening of September 17, 1853, Henry David Thoreau was searching for moose along a stream. It was his second trip to the Maine woods, and he was with Ed Hoar, his frequent traveling companion, and Joe Aitteon, an Indian guide the two had hired. "The harvest moon had just risen, and its level rays began to light up the forest," Thoreau wrote. "The lofty, spiring tops of the spruce and fir were very black against the sky, and more distinct than by day," so that "the beauty of the scene, as the moon rose above the forest, . . . would not be easy to describe." The forest in the moonlight was a place of great beauty for Thoreau on that evening, just as many a wilderness hiker hopes to find it today, often with the words of Thoreau in his mind or perhaps a copy of *Walden* in a backpack.

But then, Thoreau "suddenly saw the light and heard the crackling of a fire on the bank, and discovered the camp of . . . two explorers," who were standing at the fire "in their red shirts, and talking aloud of the adventures and profits of the day." Today, we call such men timber cruisers or exploration loggers. These same men had traveled briefly with Thoreau on their way into the woods, and so they shared a common beginning, each on his own quest, the explorers focused on physical phenomena, finding timber for commerce, Thoreau on a broader search. "I often wished since that I was with them," Thoreau noted in his journal. He described their work in a "search for timber" as climbing hills and tall trees and exploring streams to see which could be used to float logs down to the mills. They "spend five or six weeks in the woods, they two alone," he wrote, "a hundred miles or more from any town, roaming about, and sleeping on the ground where night overtakes them,—depending chiefly on the provisions they carry with them." He called their life "solitary and adventurous" and "nearest to that of the trapper of the West, perhaps," more like the experiences of Edwin Bryant when he crossed the continent to California. Their work was "with a gun as well as an axe" and "far within a wilderness." Thoreau seemed to desire to experience their life; cruising timber seemed to offer adventure beyond what he might otherwise achieve. Thoreau expressed no objection to the fact that the red-shirted men's work was to find, cut, and transport trees.

This acceptance of timber harvesting and admiration for the activities of loggers were not isolated experiences for Thoreau. During his first trip to the Maine woods, he wrote that "driving logs must be an exciting as well as arduous and dangerous business." Here, Thoreau connected danger with excitement, just as he found timber cruising adventurous. His admiration for the exploration loggers extended to others who worked outdoors with nature's resources, whether in the woods or on the water.

Thoreau clearly loved the woods and what he considered wilderness, but he also enjoyed the luxury of an extravagant campfire. Just the night before, on September 16, he and his companions had built a big fire at their camp. Thoreau wrote, "We burned as much wood that night as would, with economy and an air-tight stove, last a poor family in one of our cities all winter." Thoreau found this "very agreeable" in part because it was "independent." Once again, it was the adventurousness of the activity, along with the sense of independence, self-sufficiency, and contact with nature it afforded, that Thoreau admired. He enjoyed an extravagant campfire that might seem to many modern people wasteful and destructive, an unsustainable act not characteristic of an environmentalist.

And so we discover in Thoreau's writing what appears, from our twenty-first-century perspective, to be a strange duality: a love for the beauty of forested wilderness as well as a strong desire to participate in its logging and its use—even firsthand, even extravagantly. How could Thoreau, the father of modern environmentalism, approve of the exploration loggers and, even worse, envy them and desire to emulate them? How could any person searching for a spiritual relationship with forested nature approve of the cutting of that forest?

Perhaps the contradiction is only apparent, arising from our modern attitudes about the connection between spiritual and physical qualities of the forest and of nature in general. Part of the resolution of the apparent

contradiction is revealed in what Thoreau wrote in *Walden:* "Fishermen, hunters, woodchoppers, and others, spending their lives in the fields and woods, in a peculiar sense a part of Nature themselves, are often in a more favorable mood for observing her than philosophers or poets, who approach her with expectation." By *expectation,* I believe, Thoreau meant preconceived notions. He understood that the work of fishermen, hunters, and woodchoppers taught them about nature in a concrete, tangible way, apart from ideologies and mythologies. "Very few men can speak of Nature with any truth. They overstep her modesty and confer no favor," Thoreau wrote. "The surliness with which the woodchopper speaks of his woods, handling them as indifferently as his axe, is better than the mealy-mouthed enthusiasm of the lover of Nature."

Thoreau valued direct experience, rejected empty verbiage. Today, we could add more professionals to Thoreau's list of those who gain knowledge from direct experience, including wildlife managers and restoration ecologists, who also work directly in the woods, fields, and streams. Thus, it was the *knowledge* of nature that Thoreau valued, seemingly almost more than nature's *existence.*

Regarding experience, knowledge, and contact, Thoreau could feel that he was in wilderness when he was on land that had been partially logged, whose streams were dammed, and some of whose lakes were artificial creations of those dams. When he later tried to find his way through Umbazooksus Swamp, during his third trip to the Maine woods, he found that he was "soon confused by numerous logging-paths" made by lumbermen who had converted the area from "what was called, twenty years previous to his visit, the best timber land in the state" and "covered with the greatest abundance of pine." On that visit, Thoreau found pine "an uncommon tree." He was well aware of the human effect on the woods. Yet a few paragraphs later, he describes going through a "wilderness of the grimmest character" densely occupied by cedar trees. The proximity of a selectively logged area to a dense stand of cedar did not affect Thoreau's sense that the latter *was* a wilderness. The effects of people or their occasional presence did not destroy his *sense of being within a wilderness.* Logging per se did not interfere with Thoreau's appreciation of the spiritual qualities of forested nature, as long as the cutting was not so large in area or so severe as to disallow any sense of contact with the forest.

Thoreau did not see commerce and economic development as functioning in opposition to nature. At the beginning of this second trip to the Maine woods, he traveled up Moosehead Lake on a small steamboat called *The Captain King* and observed, "These beginnings of commerce on a lake in the wilderness are very interesting." The scenery from the steamboat was "wild," he wrote, and "varied and interesting." When the boat reached the northern end of the lake, it docked at "a long pier projecting from the northern wilderness." From the pier, the passengers and their goods were then transported by a "rude log railway," pulled by an ox and a horse, to a cabin where they stayed overnight. Spruce and fir "crowded to the tracks," and "sometimes an evergreen just fallen lay across the track with its rich burden of cones, looking, still, fuller of life than our trees in the most favorable positions." In this trammeled land, Thoreau commented that one "did not expect to find such *spruce* trees in the wild woods." It was "through such a front yard did we enter that wilderness," he wrote.

At another time during Thoreau's second trip to the Maine woods, he came across a stack of hay that had been put there for a lumberman to use during the winter; he also mentioned several places where there was "a slight frame of a camp" for lumbermen or hunters. Yet he still referred to the land as wilderness. The opening up of the forests to commerce seemed simply interesting to Thoreau, not a terrible destruction of nature that would prevent him from appreciating the forest, lakes, and streams in either a physical or a spiritual way.

One might guess that Thoreau's sensibilities were not offended by logging and other commerce because he was unable to imagine their long-term effects on a large scale. But this was not the case. Thoreau recognized the eventual effects of logging on Maine's forests and associated manipulation of rivers. On Tuesday, July 28, 1857, during his third and last trip to the Maine woods, Thoreau, Ed Hoar, and their Indian guide, Joe Polis, crossed Chamberlain Lake, which was near Umbazooksus Swamp and which he called "another noble lake." After canoeing twelve miles across the lake, Thoreau observed that the outlet on the far shore was typical of those of many lakes in the region, being an "undistinguished point" where the water "trickles in or out through the uninterrupted forest, almost as through a sponge." This outlet was dammed, however, and the three men carried their canoe over the dam. A quarter of a mile beyond, they found a second dam. The purpose of the dams, Thoreau noted, was to make it possible to change the direction of flow of the St. John River at will so that its headwaters flowed into Bangor, permitting easy transport of logs. "They have thus dammed all the larger lakes, raising their broad surfaces many feet," he wrote. Moosehead Lake, the largest of these lakes, was altered in this way. Thus, the people Thoreau referred to as "they" were "turning the forces of nature against herself, that they might float their spoils out of the country." But this meant, he wrote, that they would "rapidly run out of these immense forests."

In describing his second trip to the Maine woods, Thoreau quoted an expert whom he respected: "Humboldt has written an interesting chapter on the primitive forest, but no one has yet described for me the difference between that wild forest which once occupied our oldest townships and the tame one which I find there to-day." And so he proceeded to make the comparison himself. The "wild, damp, and shaggy look are gone," he wrote, when the "countless fallen and decaying trees are gone." The thick coats of moss also would be gone, and the earth would be "comparatively bare and smooth and dry."

But it was not the logging of forests in itself that bothered Thoreau. Rather, it was the permanent loss of resources *without any subsequent use that had human benefit* that concerned him. He criticized the loss of the sources of timber for use by cities. "Maine will soon be as deforested as Massachusetts," he wrote during his second trip. He cited descriptions by early Massachusetts settlers of well-wooded land that was in Thoreau's day no longer forested, having been cleared and settled. The wharves of Boston were built of timber cut in Nahant, Massachusetts, which had become a resort town by Thoreau's time. Considering that it had by that time become "difficult to make a tree grow" in Nahant, Thoreau wrote that "we shall be obliged to import the timber for the last" and "we shall be reduced to [gnawing] the very crust of the Earth for nutriment." He was criticizing the lack of what we today call sustainable forestry practices—the removal of timber in a way that will enable production of that resource to continue in the future. He was criticizing not the use of forest resources per se but bad methods of harvest and poor utilization.

Once the land was cleared of timber, the "they" then "[left] the bears to watch the decaying dams. Not clearing nor cultivating the land, nor making roads, nor building houses, but leaving it a wilderness as they found it." If logging cleared the land, people should at least civilize and use it. To Thoreau, cutover but unimproved land was still wilderness, whereas to us it is no longer wilderness in the sense of land untrammeled, untouched by human hands. Here, Thoreau's idea of wilderness is perhaps closer to the biblical sense of a desert, not only empty of vegetation, of nonhuman life, but also empty of civilization. Such a landscape is portrayed in the famous painting *St. Jerome in the Wilderness.* In the foreground is the saint in the desert; in the background, not far away, are the structures of civilization. In this usage, *wilderness* represents a lack of civilization in the sense of removal of a person from what is valued and safe. Wilderness is a challenge to one's soul.

Today, we rarely speak of wilderness in that biblical sense. The closest we come, as a rule, is the way some mountain climbers, spelunkers, scuba divers, Arctic explorers, sailors who venture alone in around-the-world races, and wilderness hikers see their experiences primarily as tests of endurance in which they confront nature alone and conquer both themselves—their fears—and nature by surviving. But here, the emphasis is placed on the physical test, whereas in the biblical usage, the harshness of the wilderness, including its separation of a person from civilization, was a spiritual test.

But let us return to the apparent duality in Thoreau's life and work: How could he, the father of environmentalism, the protagonist for wildness, have tolerated logging and admired loggers to the extent of perhaps wanting to join them? How could he have been so disrespectful of forest resources as to luxuriate in a huge, wasteful campfire? Where was his sense of practicing conservation simply for the sake of the forest and its creatures, beyond the concerns of human beings, material and spiritual? If he mourned the loss of spiritual inspiration provided by forests, how could he have written so positively about clearings and settlements?

It seems to me that to resolve the paradox of Thoreau's reactions, however strange those reactions may seem to us, we must understand that for him, *wildness* was a *spiritual state* arising from the relationship between a person and nature, whereas *wilderness* was land or water unused at present by people and thus a *physical state of nature.* "The Anglo-American can indeed cut down, and grub up all this waving forest, and make a stump speech, and vote for Buchanan on its ruins," Thoreau wrote, "but he cannot converse with the spirit of the tree he fells, he cannot read the poetry and mythology which retire as he advances." In this view, when loggers destroy a forest, they lose in two ways: they fail to conserve the land for other productive human uses, such as farms and settlements, and they lose an opportunity to commune with nature. Loss of the human benefit of a spiritual connection with nature was what Thoreau mourned, but he himself was able to attain this connection on partially logged land.

Wildness meant very much to Thoreau, to the extent that elsewhere he wrote: "I caught a glimpse of a woodchuck stealing across my path, and felt a strange thrill of savage delight, and was strongly tempted to seize and devour him raw; not that I was hungry then, except for that wildness he represented."

Wilderness, the physical entity, was a place, then, where a person could experience wildness; if it was destroyed for this use, Thoreau believed, it should be converted to other humanly productive uses. Compare this position with the popular belief of our time. Today, wilderness is set forth by its proponents as having an intrinsic—that is to say, intangible—value in the sense of possessing a moral right to exist and having great importance to the workings of the biosphere and the persistence of all life. Wilderness is today defined as a place without people and free of their influence. Yet for Thoreau, people were as much a part of wilderness as they were a part of civilization. The importance of wildness was in its spiritual meaning to human beings. The modern conception of wilderness separates people from nature. For Thoreau, such separation would destroy one's ability to appreciate nature's spiritual and inspirational values.

On that night in the Maine woods when he was hunting moose and found the exploration loggers, in the darkness behind him was a forested wilderness—influenced and affected by human beings but nonetheless Thoreau's wilderness. In the firelight of the timber explorers' camp, Thoreau saw the adventure, the experience, that was contact, contact, contact with nature. In the Maine woods, Thoreau confronted his material and spiritual needs and desires for nature—confronted them, separated them as concepts, and then determined for himself how they were interconnected. As a result, he admired those who had direct experience with nature for their knowledge of nature, both physical and spiritual. He respected the use of living resources, including forests, as long as these uses either were sustainable or led to other uses of the land that were of human benefit. And, unsaid in the passages quoted here but evident in his writings quoted earlier, Thoreau was satisfied as long as there was the equivalent of the swamp on the edge of town or the forest's soil beneath the village—as long as there was, nearby, a way for a person to find his connection with nature and to make that connection often. It is a different perspective from today's. Its benefit is that it brings people into contact with nature and leads them to respect both wilderness and civilization. It is not a justification for eliminating large areas of wilderness; rather, it is a justification for renewing an emphasis on naturalistic countryside near urban and suburban areas. The question that it poses for us is whether Thoreau's approach, which builds from his approach to knowledge of nature in both its spiritual and physical aspects, is appropriate for our time.

Finding Salmon on the Merrimack

"Science itself is now the only field through which the dimension of mythology can be again revealed."

Joseph Campbell, *The Masks of God*

"It is the marriage of the soul with Nature that makes the intellect fruitful, and gives birth to imagination."

Henry David Thoreau, *Journal*

On Sunday, September 1, 1839, Henry David Thoreau and his brother, John, rowed on the Merrimack River. It was the second day of their trip in the boat that the two had built. Thoreau observed that this river had banks "generally steep and high" and, interestingly, that the river was "probably wider than it was formerly, in many places, owing to the trees having been cut down, and the consequent wasting away of its banks." He saw that the river descended "over a succession of natural dams, flowing through still uncut forests, by inexhaustible ledges of granite, where it has been offering its privileges in vain for ages, until at last the Yankee race came to *improve* them—a silver cascade which falls all the way from the White Mountains to the sea."

Because of the dams, salmon were in decline. "Behind a city on each successive plateau," Thoreau wrote, there was "a busy colony of human beaver around every fall. Not to mention Newburyport and Haverhill, see

Lawrence, and Lowell, and Nashua, and Manchester, and Concord, gleaming one above the other. Shad and alewives are taken here in their season, but salmon, though at one time more numerous than shad, are now more rare. Locks and dams have proved more or less destructive to the fisheries."

In the 1790s, dams and canals changed the nature of the Merrimack, but reports indicate that even as late as 1805, fishermen could get twenty and sometimes forty salmon per day. By Thoreau's time, much of the fish population was already gone. Thoreau knew that the Merrimack was much altered from its condition before European settlement—especially regarding undesirable changes in the habitats and numbers of native fish. But he still found the river inspiring. The specific, material changes in the Merrimack were not enough to destroy his pleasure and sense of wonder as he rowed on the river. He felt a spiritual connection with nature there, in spite of the human influence. Thoreau was aware of other such changes taking place, including a reduction in the abundance of wildlife. In his journal, he mentioned the declining number of passenger pigeons.

Perhaps Thoreau's recognition of human alterations was limited to New England. If so, would he have arrived at different conclusions about civilization and nature, about the human spirit and nature—if he were confronted with modern environmental problems, including the huge human population, worldwide alterations of nature, accelerating extinctions of species, and global environmental pollution? And if Thoreau were alive today, would he hold different views about forestry and foresters from those he expressed during his lifetime? Although these questions arise quite naturally, we cannot resolve them completely; we can only speculate about them in hindsight. A more productive question is, Where does Thoreau's approach to knowledge of nature, both spiritual and physical, lead us today?

As we desperately try to restore the environment to previous conditions, can the new, altered nature still inspire us? Must nature be perfectly virgin and original, or can we find a connection between our feelings and a modified—a naturalistic—set of physical surroundings or even a natural environment we are trying to restore? If some modified forms of nature still leave us with a sense of what is natural, what are the limits on such modification? Have we already gone too far, or will some writer in the future inspire readers by describing a nature even more altered than our own?

In confronting these questions, it is helpful to consider what was known about the natural history of North America during Thoreau's lifetime. How well was nature in the large observed and known in a physical sense?

Important scientific exploration of North America had begun before Thoreau. Some was ongoing during his lifetime. Alexander von Humboldt had completed his explorations of Central and South America by 1804, thirteen years before Thoreau was born. Humboldt published his important geographic work in a series of volumes from the time Thoreau lived at Walden Pond until Thoreau's death—between 1845 and 1862.

Meriwether Lewis and William Clark made the first major natural history expedition across the western United States, from St. Louis to the mouth of the Columbia River, between 1804 and 1806. They described the conditions of plant and animal life and the geology of the American West, sending back samples of flora and fauna, including live prairie dogs. They described 122 new species and provided the first formal natural history descriptions of many animals and plants, including the mule deer and the badger. However, although Lewis and Clark completed their exploration of the American West eleven years before Thoreau was born, their journals were not widely available until after Thoreau's death, and it is unlikely that Thoreau ever saw them.

Pehr Kalm, a Swedish botanist, came to eastern North America in 1748 to collect and describe North American vegetation for Carolus Linnaeus. Between 1749 and 1750, Kalm traveled from Philadelphia to Montreal, collecting plants; later, he wrote about the natural history of the countryside and commented on social conditions and beliefs about nature. William Bartram described the vegetation of the southeastern United States, primarily Florida, in *Travels through North and South Carolina, Georgia, East and West Florida,* published in 1791. As is Kalm's journal, Bartram's book is a personal narrative that includes his observations of the plants and other aspects of natural history. Thoreau's *The Maine Woods* is somewhat in the genre of these two works.

John James Audubon, who was born in Haiti in 1785 and died in New York City in 1851, began drawing the birds of North America in 1803, fourteen years before Thoreau was born. His drawings were first published as hand-colored prints in the four-volume *Birds of America* between 1827 and 1838, when Thoreau was young. Other explorations of the animals, plants, soils, minerals, and bedrock of North America were under way during

Thoreau's lifetime. It was in this sense a period of discovery and recording of the countryside's condition.

The discoveries of new species and new landscapes had a strong effect on ideas about nature, both scientific and religious. In a religious context, the discoveries of species revived the ancient questions of why there are so many species on the Earth and how they fit within the great chain of being—that is, the purpose or role of each species. Among the most important eighteenth-century writers on this issue was William Derham, who in 1798 published *Physico-Theology: Or, a Demonstration of the Being and Attributes of God, from His Works of Creation.* He discussed discoveries of new plant and animal species by European explorers and naturalists, discoveries that had begun during the age of exploration and were continuing in his time. His goal was to place these discoveries within the context of Christianity. He struggled with several issues, including the question, With so many kinds of creatures living on the Earth, each kind having a great capacity for reproduction, what prevents the world from becoming overpopulated and falling into disorder? "The whole surface of our globe can afford room and support only to such a number of all sorts of creatures," he wrote. These creatures could, by their "doubling, trebling, or any other multiplication of their kind," increase to the point that "they must starve, or devour [one] another." That this did not occur, Derham held, was evidence of divine order and purpose. The keeping of the balance of nature, he wrote, "is manifestly a work of the divine wisdom and providence."

Derham also tried to deal with the more difficult problem of why vicious predators existed on an Earth made by a perfect God. The newly discovered Peruvian condor stimulated Derham's thoughts. He called the condor the "most pernicious of birds," describing it as "a fowl of that magnitude, strength and appetite, as to seize not only on the sheep and the lesser cattle, but even the larger beasts, yea the very children, too." Condors seemed to him the rarest of animals, "being seldom seen, or only one, or a few in large countries; enough to keep up the species; but not to overcharge the world."

Derham gave many other examples of predators, all of which were rare in comparison with their prey. He wrote that this was a "very remarkable act of the Divine providence, that useful creatures are produced in great plenty," whereas "creatures less useful, or by their voracity pernicious, have commonly fewer young, or do seldomer bring forth." This, then, was the mechanism that maintained the "balance of the animal world," which was "throughout all ages, kept even." He continued, "By a curious harmony and a just proportion between the increase of all animals and the length of their lives, the world is through all ages well, but not over-stored." This was a divine order.

Derham's conclusions repeated explanations given by earlier Christian writers, who in turn echoed the explanations of classical Greek and Roman philosophers. Derham was restating what has become known as the "design argument." This is the argument that the universe, the Earth, and life on the Earth are too complex to have come about on their own—out of chaos, so to speak—and therefore must have had a creator, who designed and constructed all that we call nature.

The use of natural resources in and before the eighteenth century also led to discussions about negative effects of human beings on nature. Jacques-Henri Bernardin de Saint-Pierre wrote in the eighteenth century about "noxious insects which prey upon our fruits, our corn, nay our persons." He regarded these unpleasant and undesirable occurrences as a result of human activities, for "if snails, maybugs, caterpillars, and locusts, ravage our plains, it is because we destroy the birds of our grove which live upon them." Or, he wrote, by introducing the trees of one country into another, "we have transported with them the eggs of those insects which they nourish, without importing, likewise, the birds of the same climate which destroy them." He believed that a natural balance existed within each country, a balance that could be upset by human actions. Every country had birds "peculiar to itself, for the preservation of its plants." Birds that fed on insects had their role in the divine order, which was being upset by human beings. From Derham and Bernardin de Saint-Pierre we learn that before Thoreau, others were concerned with issues that today we call environmental: the "green" issues of the loss of species and the effect of that loss on the physical and spiritual state of humankind. Thoreau did not live in an environmentalist's vacuum; rather, he lived in a time when rapid discoveries about nature were raising questions similar to those with which we struggle today.

But one can still ask whether the *quantitative* human effect on nature has become so large as to force a rejection of an approach to nature like Thoreau's. The great increase in the human population, underlying all modern environmental problems, comes immediately to mind. The human population of the world in 1800 was about 900 million, less than 20 percent of the population today. The population of the United States in 1810 was slightly more than 7 million, not much larger than the present population of greater metropolitan New York City or of the Los Angeles Basin. In 1820, the U.S. census recorded fewer than 10 million people. At the time of Thoreau's birth, the population of the United States was therefore between 8 million and 9 million. By the time of his death, the U.S. population had grown to more than 31 million, about 12 percent of the 1990 population.

Before Thoreau, Thomas Robert Malthus had recognized and made widely known the potential problems of human overpopulation. His famous *Essay on the Principle of Population,* first published in 1798, remains perhaps the most eloquent statement of the problem ever made. Malthus began with two premises: food is necessary for people to survive, and, as he stated it, "passion between the sexes is necessary and will remain nearly in its present state" so that children will continue to be born. As a result, the power of population growth, he wrote, is "indefinitely greater than the power of the Earth to produce subsistence."

In other words, Malthus reasoned that an ever-growing human population would be impossible to sustain with a finite resource base. His projections of the ultimate fate of humankind were dire, as cynical and pessimistic as any to be found today. This power of population growth is so great, he wrote, that "premature death must in some shape or other visit the human race. The vices of mankind are active and able ministers of depopulation, but should they fail, sickly seasons, epidemics, pestilence and plague, advance in terrific array, and sweep off their thousands and ten thousands." And worst of all, should these fail, "gigantic famine stalks in the rear, and with one mighty blow, levels the population with the food of the world." Not only was Malthus's statement powerful rhetoric, it was influential among the greatest thinkers of Thoreau's time. It was a primary factor in the reasoning of Charles Darwin that led him to the theory of biological evolution. The problem of the future of the human population was well understood when Thoreau was writing.

The Louisiana Purchase took place only fourteen years before Thoreau's birth. As a political entity, the United States consisted of states east of the Mississippi River and a vast territory to the west. The pressure of human populations on natural resources was obviously much lower on the continent at that time, although the pressure may have been severe at certain locations. One might think that this great difference in human population size would be enough to make Thoreau's perception of nature entirely different from ours, but the area he knew best, the lands around Boston, had long ago been cleared and converted to various uses, and alterations of nature were locally intense. As Thoreau described in *The Maine Woods,* the clearing of forests near Boston had greatly changed that landscape in comparison with the landscape of Maine.

Cape Cod, now intensely populated and heavily used for tourism, was rural countryside in Thoreau's day. Pilgrims and other early settlers had much altered the Cape, however, as Thoreau himself noted in *Cape Cod.* From the vistas around Concord, the ones he saw most often, human effects on the landscape were great and not inconsistent with what we see today. The great differences between Thoreau's time and ours include results of the invention and use of thousands of artificial chemicals; human effects on the oceans, including marine fisheries; and evidence suggesting global environmental consequences of human activities.

The wild living resources of the United States were in a vastly different state in Thoreau's time. The history of the relationship between these resources and Western civilization in North America can be divided roughly into five periods. The first, the period of discovery, began with the arrival of Europeans. During this period, the geographic ranges of many species were reduced, but the effect on entire species, although significant in certain cases, was small overall in comparison with the events that followed.

The second period was a time of intense exploitation of the continent's wild living resources. The timing of this exploitation varied somewhat with species, resources, and geography, but in general it lasted from the early 1800s to the beginning of World War I. During this period, resources were extracted without much thought about their ability to persist or, if there was some thought about it, little care or attention.

Thoreau lived during this second period. Some of the exploitation took place in and from Massachusetts and elsewhere in New England, which was a major center for industry and manufacturing. New Englanders were some of the prime movers in the extraction of certain biological resources. Among these were the great pelagic whaling fleets that shipped out of New Bedford, Massachusetts. The New Bedford whalers did not begin to ship in great numbers until the 1840s. The demise of the bowhead whale brought about by these ships was just starting when Thoreau sat and philosophized in his windy Walden hut some miles—but not so many—from New Bedford.

Between 1840 and 1920, with great precision and economic frugality, whalers hunted the mighty bowhead whale. They took approximately 10,000 of these huge creatures between 1840 and 1850 and another 10,000 in the next decade, eventually reducing the overall population from approximately 50,000 to about 3,000. Thoreau wrote in support of the economic harvesting of small whales, which he saw being hunted during his trips to Cape Cod, and never expressed concern about the whaling fleets based in New Bedford.

As Thoreau rowed on Walden Pond, sea otters played and elephant seals dived deep off California's coast in great numbers. Trade in the fur of sea otters, which had started in the late eighteenth century, was well under way. The great fur trader John Jacob Astor founded Fort Astoria at the mouth of the Columbia River in Oregon in 1811, creating the first permanent settlement along the northern Pacific coast of what is now the United States and precipitating the exploitation of furbearing animals of the Pacific Northwest, which led to the near demise of the sea otter and elephant seal.

Bison were plentiful and roamed the Great Plains throughout Thoreau's lifetime, their herds sometimes so densely packed that they obstructed passenger trains. It was not until the 1880s, two decades after Thoreau's death, that Buffalo Bill, Wild Bill Hickok, and other hunters killed off most of the bison. After this, a few concerned citizens took the remaining few hundred into protective custody.

Prairies, which once covered more land area in the United States than any other vegetation type, were essentially eliminated and replaced by agriculture, except for a few scattered acres. Salmon were almost lost to the eastern United States, and their abundance and distribution were greatly decreased in the West.

Professions of formal management for forests and fisheries lay ahead, at the beginning of the twentieth century. These were part of a third period in the relationship between Western civilization and wild living resources—a period of awakening in conservation that began between the turn of the twentieth century and the end of World War I and lasted until the 1960s. Exploitation continued but was accompanied by the first attempts at professional management of forests, fisheries, and wildlife and the beginnings of a national conservation movement, a sociopolitical movement with intellectual antecedents in the nineteenth century. During this third period, resource management had a single goal: to maximize the production of individual resources for harvest, as if wild living resources were farm crops or could be treated as such. The United States established the Forest Service as part of the U.S. Department of Agriculture. Yale University founded a Forestry School, the first university institution devoted to training professionals for the new job of maximizing the economic growth of trees. It was the era of Gifford Pinchot, head of the new Forest Service and founder of the Yale Forestry School.

A fourth period, that of public environmentalism, began in the 1960s. Public awareness of environmental degradation and the decline of wild living resources grew, spread, and became established as a sociopolitical movement. Environmentalism developed rapidly, accompanied by the passage of landmark legislation to protect and conserve the environment, such as the Water Quality Improvement Act of 1970, the Marine Mammal Protection Act of 1972, and the Endangered Species Act of 1973.

We are today, perhaps, in the midst of a fifth, unnamed period, one with socially schizophrenic tendencies. On the one hand, people are engaging in major constructive actions to conserve the environment. A large suite of corporations and government agencies seek to solve environmental problems through applied science, computer methodologies, civil and chemical engineering, and some ecological concepts. Numerous corporations in all sectors are integrating environmental concerns into their standard procedures, and many have environmental officers. There is even a professional journal titled *Corporate Environmental Strategy.* The National Research Council has conducted a study of corporations and the environment. Major environmental organizations, national, regional, and local, have had a significant influence on policy and on the conservation of landscapes. Most people in the United States consider themselves environmentalists; the country has cleaner air and water; and the Endangered Species Act is a powerful law that has had a major effect on procedures.

On the other hand, there are so many people in the world today—approximately 6 billion—and so many changes are occurring in nature that the sheer number of people and force of their effects may have led to a qualitative difference in the relationship between people and nature. Even on the Merrimack River, the environment continues to experience problems. Salmon were reintroduced there in 1975, and hatcheries began to release more than a million fry each year, yet of the million fry released, fewer than two hundred adults return. Conflicts between nature and civilization seem only to intensify.

Given the modern situation, to what extent is Thoreau's approach to nature and civilization relevant to our time? It is an open question. Do the quantitative changes lead to qualitatively different issues that require entirely different approaches, as the deep ecologists suggest? One could argue that viewed from Thoreau's more limited perspective, based on his experiences in eastern Massachusetts and his travels to Maine and Cape Cod, the changes in New England during his time were as intense as those we find worldwide today. We could, then, attempt to take a constructive approach from the methods he used.

Toward the end of his trip on the Concord and Merrimack Rivers, Thoreau wrote, "When compelled by a shower to take shelter under a tree, we may improve that opportunity for a more minute inspection of some of nature's works." He and his brother reached the northernmost limit of their boat travel on the Merrimack on a rainy day. It was at a point where they would have had to portage, and the boat was too heavy to be carried around rapids. Rather than turn back immediately, Thoreau and his brother chose to walk along the river's bank, "feeling our way with a stick through the showery and foggy day, and climbing over the slippery logs to our path with as much pleasure and buoyancy as in the brightest sunshine; scenting the fragrance of the pines and the wet clay under our feet, and cheered by the tones of invisible waterfalls."

Thoreau was suggesting that we can take constructive action to help ourselves by beginning with detailed observations. "I have stood under a tree in the woods half a day at a time during heavy rain in the summer, happily and profitably there prying with microscopic eye into the crevices of the bark or the leaves or the fungi at my feet," he wrote. Was this the marginal behavior of an abnormal personality or a guide to our future?

We today could give up on Western civilization, or on civilization completely, as the deep ecologists suggest, or we could simply dismiss environmental issues, as proponents of the wise use movement suggest. But most people do not want to do either: as mentioned earlier, recent public opinion polls suggest that people who are concerned about nature are also most likely to be concerned about people and civilization. But for Western civilization to continue, we must find a path that allows nature and civilization to be compatible.

I suggest that Thoreau's approach to knowledge of nature remains a key to sustaining nature and civilization together. Thoreau's ability to recognize the importance of both, to distinguish between them and yet link them together, is as important today as it was in his time—perhaps even more so. The components of Thoreau's careful approach to knowledge—his emphasis on direct, detailed personal observation, letting generalizations arise from the observed details; his use of local experiential expertise as a starting point for new ideas; his similar use of professional expertise, both written and spoken; his objective approach to observing nature; his use of quantitative measurements and his openness to new ideas, perhaps influenced by his experiences as a surveyor and inventor—are keys to what we need to do.

I believe that through Thoreau's understanding of human spiritual as well as physical needs and desires for nature, we can seek to maintain both civilization and nature, and maintain a civilization that continues the rationality, democracy, innovation, and creativity of modern Western civilization. We can put Thoreau's approach to the test in current environmental issues. Our path may appear more difficult because of the events that have taken place, but his approach remains viable. Perhaps it will be helpful, or perhaps it will work no better for all of nature and all of civilization than a strange examination of a piece of bark or a leaf in the rain in a woodland summer. Personally, I believe we, Western civilization, and nature will all benefit from a restoration of Thoreau's approach.

Chapter 12

Putting Forests on the Ballot

"Men and nature must work hand in hand. The throwing out of balance of the resources of nature throws out of balance also the lives of men."
Franklin D. Roosevelt

"Our minds anywhere, when left to themselves, are always thus busily drawing conclusions from false premises."
Henry David Thoreau, *The Maine Woods*

FOR MANY DECADES after Henry David Thoreau saw the red-shirted timber explorers by their campfire, few people paid much attention to the Maine woods. The woods were far away and out of sight. Tourists went to the Maine coast, Mount Desert Island, and Penobscot Bay to eat lobster, go sailing, and watch the clouds soar over the waters. A few hikers pursued the Appalachian Trail to its end at the summit of Mount Katahdin, some taking a side trail to follow the path that Thoreau discovered on his first ascent. The occasional canoeist passed along the streams and waters.

But then came the 1990s and an intense public interest worldwide in forests as nature, in forests for their biological diversity, in forests for their role in global climate change, and in ancient forests as a kind of cathedral of nature. Concern with forests rose to the top of the list of environmental issues. Clear-cutting became one of the most contentious such issues, perhaps the touchstone of environmental controversies, not just in Maine but all over the world. The contentiousness extended to the question of whether old-growth forests are necessary and, if so, to what extent the landscape must be made up of old growth. *Old growth* is an informal term for a forest that either has existed for a long time without any kind of disturbance, human or otherwise, or has never been disturbed.

In forestry, the term *clear-cutting* means clearing the land of all its trees, but informally it also means cutting most of the trees so that the landscape appears devastated. Under the old idea of a balance of nature, it was believed that a forest grew from a clear-cut to a single, final condition, at which it would remain indefinitely. Ecologists referred to this condition as the "climax" stage of a forest. In recent years, with the recognition that forests are dynamic rather than static, the less formal term *old-growth forest* is generally used instead, even in legal documents. What is important here is that a clear-cut forest and an old-growth forest are taken by the public to represent opposite ends of the spectrum of forest change.

Recently, Greenpeace International lobbied European countries to stop importing wood and paper from Canada because the wood was harvested by clear-cutting. In 1996, environmental organizations opposed to clear-cutting scared two big timber-producing countries, Canada and Malaysia, into holding a large conference near Malaysia's capital city, Kuala Lumpur, at a cost of about $0.5 million. The conference brought together scientists and interest groups to discuss how to certify that forest products have been obtained by sustainable forestry methods.

Sustainability is no small issue. Wood is and always has been a major resource for civilization. In 1990, forests and other wooded lands—open woodlands such as the ponderosa pine forests of the western United States, savannas such as those in East Africa—covered 40 percent of the Earth's land surface. Closed forests, in which the leaves of one tree touch those of another to form a continuous green canopy, covered 3.4 billion hectares, or 12 million square miles. Annual trade in timber for construction, paper, and pulp amounted to 1.5 million cubic meters. An equal amount probably was used locally for firewood in many regions of the world where firewood is the primary fuel source.

People recognize that forests play a key role in global environmental processes. Forests affect the storage and release of greenhouse gases. By moderating the reflectance properties of the Earth's surface and the amount of water evaporated, they influence weather and climate. The continued production of wood as a resource for primitive peoples, for advanced civilizations, and for the biosphere is important both for civilization and for the Earth's life-support system. These are utilitarian and ecological—pragmatic—concerns.

On November 2, 1996, Maine citizens voted on a Referendum 2A, known as the "Clearcutting Referendum," which would have banned clear-cutting on millions of acres of the state's forestlands, the very lands through which Thoreau had once walked and canoed. A large clear-cut forest looks like an ugly scar on the landscape, a wound from which it might seem the land would never recover. A clear-cut looks especially horrible where it is surrounded by beautiful, intact forests or where it extends from horizon to horizon. Its very existence strikes deep at the heart of the modern sensibility of what is physically beautiful and spiritually meaningful in nature. It is as if the very spirit of the woods has fallen victim to the scalpel.

Perhaps the ugliest example of massive clear-cutting I have seen was on the Olympic Peninsula in Washington State in the mid-1990s. Driving through there and using my car's odometer to gauge the extent of clear-cutting, I clocked twenty-five miles on the main north–south road where the forests had been cleared from the edge of the road to far in the distance. At another location, a large clear-cut stood just across the highway from a major tourist lodge within an old-growth stand. When I asked my forester colleagues who live in that region why the timber companies had not left even a narrow strip of old growth to hide the worst of the logging from the public—something that would seem to have been in their best interests—and why they had logged right to the border of the entrance to the tourist lodge, I was told that the loggers were so angry with environmental opposition to their work that they had adopted an in-your-face, we-will-do-as-we-please attitude. I could not verify the truth of the answer, but it was consistent with my worst fears about our society moving further from a middle ground toward extremes, and away from solutions beneficial to either civilization or nature.

On the other hand, clearings play a natural role in forests. Many tree species, for example, regenerate only in clearings. In the Maine woods, these include white pine, the premier timber tree; white birch, which is valued as a hardwood and whose bark gives great beauty to the landscape; and yellow birch, a species increasingly valued for use in furniture. Out west, Douglas fir, coast redwood, and giant sequoia require clearings to regenerate. These trees do not grow well in the shade of other trees, and their seeds cannot sprout in the leaf litter and organic material found on the floor of an old-growth forest; a seed of these western conifers that falls on a dead leaf, twig, or mound of soil rich in organic matter has no chance of germinating. Some species are specifically adapted to fire, requiring it or a close equivalent to regenerate; others do best after a storm or some other kind of land-clearing event. This is true for many species that are commercially important and aesthetically desirable.

Besides observing that forests regenerate through a process of succession, in which species succeed one another in a distinct pattern, Thoreau was well aware of the role of clearings in forest restoration. His book *Faith in a Seed* discusses the regeneration of pines and oaks in Massachusetts, noting how the shade of pines provides a suitable habitat for seeds of other species to regenerate. Thoreau understood that these stages of regeneration meant that a forest was in reality a process, not a fixed entity.

Once we recognize that a forest is a dynamic, living process rather than a simple, static structure that we want to hold still, the question of how best to harvest trees becomes a technical one, of how the process works and how we can operate in harmony with it and, if possible, help it along. However, the fact that forest clearings are ecologically desirable under certain conditions is *not* an excuse to clear-cut all the land for miles around or to use any method of forestry, no matter how destructive.

Although the word *clear-cutting* is understood to mean different things—

removal of all trees or removal of enough of them to leave a landscape that appears devastated—Maine's 1996 referendum provided a specific legal definition. It would have prevented the removal of more than one-third (by wood volume) of trees more than 4.5 inches in diameter—about as wide as a person's wrist—in any fifteen-year period on any land under the jurisdiction of the state's Land Use Regulation Commission. This is much more restrictive than the usual definition of clear-cutting. A forest two-thirds as dense as another would be seen as a dense forest by many people not trained in natural history, botany, forestry, or ecology. Once it had regrown over the scars of cut stumps, it might even appear to many people as natural and wild.

A competing measure, Referendum 2B, would have allowed clear-cutting but would have required a permit that would take into account the proximity of other clear-cuts and the total area recently clear-cut by each owner. In addition, this second measure would have established an "ecological forest reserve system." Opinions about the two referendums ran deep and strong, as expressed in the newspapers and on the World Wide Web. An organization called Common Sense for Maine Forests published an article on-line claiming that the environmental organizations that defended Referendum 2A were secretive, upper-class, and trying to intentionally trash Maine's economy. According to the Associated Press, this was "the most expensive referendum in the state's history, with paper companies spending more than $5 million to defeat it."

The conflicts persist. In the very region where Thoreau canoed, hiked, and admired red-shirted timber explorers, the debate rose once again in the late twentieth century. Must all or a large fraction of forests be left unlogged and in an old-growth condition in order to save nature—to allow all forms of life to persist and entire ecological systems to maintain themselves? Must we have only one or the other, unlogged forests or civilization? The debate over the referendums renewed my fears that without a recognized connection between civilization and wild living resources, the situation would worsen for both.

Maine voters made a clear choice: Referendum 2A was defeated, and a similar measure was defeated in 1997. The majority of people did not support a ban on clear-cutting.

Typically, the argument against clear-cutting is presented in terms of its destruction of forests in a way that prevents their recovery, their sustainability, and their support of other species—their support of biological diversity. Jamie Sayen, editor of the *Northern Forest Forum,* writing in the *Earth Island Journal,* stated that "northern Maine forest stands are in their worst condition since the retreat of the glaciers approximately 12,000 years ago" because "there is virtually no old growth or mature forest" and there are few protected areas. Here, old growth is considered not only good but the best, and perhaps the only acceptable or natural, condition of a forest.

Sayen saw this issue as a political and public relations battle between extremes. He continued that "any hope of good faith bargaining with industry was shot down by the decision of the Maine Forest Products Council and other industry heavyweights to hire the anti-environmental program firm Winner Wagner and Mendelback of San Francisco."

An argument that clear-cutting is bad because it destroys life-supporting characteristics of forests was presented by John M. Hagan III, senior scientist and director of Maine programs at the Manomet Center for Conservation Sciences, Manomet, Massachusetts. A Massachusetts resident, Hagan had become interested in the Maine woods. In an article in the *Maine Policy Review,* Hagan stated that he saw the clear-cutting issue as a "bitter debate over jobs for Maine people versus the future of the North Maine Woods."

In a kind of mea culpa for which he can only be praised for his honesty and willingness to share what he had learned, Hagan told his own story. Initially, he had been interested in conducting research on a possible connection between Maine forest practices and a decline in populations of migratory birds. He had been provided with access and an introduction to land owned by the S. W. Warren Company and taken to the site of a clear-cut. On arriving there, he had been so appalled by what he saw that he was "rendered incapable of simple conversation" and decided that he would have "no need to survey the clearcuts" for his research, on the basis that "nothing could live in what [he] had seen, at least for many years." He had formed his conclusion without quantitative evidence, letting his personal reaction to his subject determine what he intended to claim was a scientific conclusion.

Here, Hagan violated the scientific method by forming an opinion that he had no intention of testing, all the while intending to make what he believed was a scientific study. This is a common fallacy, often practiced with environmental issues. It violates the approach Thoreau took in studying Walden Pond and in developing the pencil.

Hagan's host on the S. W. Warren Company's land was research forester Carl Haag. Haag and Hagan talked about the research plan, and Haag persuaded Hagan that he should make measurements in clear-cut areas as part of doing a scientific study correctly. Haag's advice is interesting; it is consistent with Thoreau's conclusion that there is high value in the knowledge of those who have worked for a long time in the woods. Hagan had approached the woods with, as Thoreau expressed it, "expectation." However, he decided to follow Haag's advice, reasoning, "Once I had *documented* that clearcuts were avian deserts, public pressure might bring about kinder, gentler, forest practices."

Much to Hagan's surprise, however, when he began his fieldwork, he found the clear-cut areas "full of birds," including many that conservationists had thought to be threatened or endangered, such as the chestnut-sided warbler and the American kestrel. To his credit, at this point Hagan abandoned his ideological prejudices and accepted the observations that confronted him. These observations led him to conclude that the conservation of biological diversity required forests of different ages, including clear-cuts. He recognized that a forest is a process, not a structure. Scientific observation had changed his opinion. Haag believes that this realization was a "life-changing event, a catharsis," for Hagan.

Once he had recognized the need for the existence of a variety of stages in forest development, Hagan realized that the clear-cutting issue, which he had at first seen as a debate over jobs versus forest preservation, now had to be rephrased as a question: What *amount* of clear-cutting is most beneficial to biological diversity? I admire Hagan for having written an article that sets forth his mistake and puts the record right. It is courageous, honest, and helpful. His article is a breath of fresh air in this time of intensifying misunderstandings. Haag observes that his changed opinions led some environmental organizations in Maine to try to discredit Hagan for what Haag said they saw as "incorrect thinking."

The Maine referendum and the bitter, expensive debate over it show that the issues that concerned Thoreau continue to rage in modern society. The willingness of Hagan the scientist to let his initial opinion override the scientific method illustrates the depth of our feelings about forested nature and about leaving nature alone.

The clear-cutting referendums illustrate the tendency for public debates to focus on beliefs rather than on scientific evidence about the material condition of forests and how the forests can and must be maintained. But these phenomenological questions stir up intensely emotional debates because underlying them are deeply held spiritual attitudes toward forests and civilization. Hagan's initial approach to his research exemplifies the depth of these concerns with the intangible and the tangible, the same duality of concerns that occupied Thoreau. It clearly illustrates how aesthetic, spiritual, and religious beliefs affect people's approaches to physical phenomena.

Forests have always held spiritual meaning for people. Sometimes the spiritual aspect concerned contact with mysterious forces that were believed to rule the world and thereby determine nature's principles: seeking the interior of a forest was one way to seek a spiritual connection with these forces. Sometimes the spiritual aspect concerned fear: fear of the dark woods and its silent, often dangerous creatures, real and imagined.

The rapid depletion of forests in recent decades continues a process as old as civilization. As John Perlin describes in his book *A Forest Journey,* since the time of Babylon, civilizations have depended on forests but have cleared them faster than they could regenerate. One by one, civilizations have used up their forest resources and been forced to seek these resources at greater and greater distances.

As a result, people have held on to a curious and important contradiction about forests. On the one hand, civilizations have depended on forest products; on the other, people have generally feared the darkness of forests. Clearing the woods brought in light and a sense of safety. Perlin's book opens by recounting the Mesopotamian epic of Gilgamesh, who became a hero by being brave enough to go into the dark forests and clear them. Similarly, in the Anglo-Saxon epic poem *Beowulf,* the hero leaves the warmth of the hearth to go out into a dark, cold world where he becomes a hero by slaying a monster, symbolically conquering wild nature and taming it.

Joseph Conrad's famous novella *Heart of Darkness* is a journey into a tropical forest that is ancient, uncut, dark, and the source of primitive, dangerous, and frightening qualities of the inner spirit of a human being. It is both tangibly and intangibly dark and dangerous. It is dangerous in the sense of evoking an internal human primitiveness—the uncivilized aspect of human nature.

Conrad's hero and alter ego, Marlowe, is a man of great experience and intellectual reflection in confronting nature—nature that we would call wilderness—on the ocean and in forests. He is reminiscent of the timber cruisers Thoreau met in the Maine woods and admired. But Marlowe, like Thoreau, had an intellectual curiosity in addition to the rugged outdoor qualities of the timber cruisers. He was driven not only to experience "raw" nature but also to understand it and, through that experience, to understand the place of people within nature. He sought both to experience nature externally, through his senses, and to use the results of those sensations to achieve a deeper level of understanding.

Wild nature in *Heart of Darkness* is not a positive place. It corrupts people, destroying their ability to work, to concentrate, to create, to be civilized. Those who survive it do so either by holding on to outward practices and trappings of their former lives in a way that seems absurd in the wilderness or by succumbing to the corrupting influences of wilderness. Conrad saw this kind of experience as having taken place throughout the history of Western civilization. He had Marlowe describe the imaginary first meeting of Romans and ancient Britons, the latter at the time a primitive people who painted their faces blue and lived in dark forests.

Marlowe's trip into the "heart of darkness" is simultaneously a physical trip into an old forest that is just beginning to be affected by modern Western technology and an introspective journey into his own ancient, primitive spirit. Marlowe views a virgin forest in the classical way: uncivilized and uncivilizing, dangerous and destructive.

Thus, from Gilgamesh to Conrad, ancient forests have challenged the human spirit. These past perceptions of wilderness suggest that Western society of the late twentieth century created a unique perception of the value of forests, especially old-growth forests. Today, they are valued fundamentally for spiritual reasons that are taken to be positive. Wilderness is a place for recreation and renewal, uplift and inspiration. Yet few people visit these wilderness places, and among those who do, even fewer understand how to integrate a spiritual relationship and a scientific understanding of forests. We have created legally designated wilderness areas whose dynamic properties few of us understand or care to understand but which we believe are important because of their supposedly static physical structure. We accept the idea that the structure of a primeval forest is profoundly important to the human spirit, yet few of us test this idea directly, in the way Thoreau tested himself on Mount Katahdin or in the wetlands of the Maine woods.

In a forested landscape, a variety of biological structural types is present at any time in different locales: clearings and old growth, areas of high and low species diversity. There is a natural range of conditions in any given location and a range of sets of conditions on any landscape. These stages in forest development are a necessary feature of the landscape; we cannot disregard them, clear-cutting all the land or ignoring the requirements for sustaining natural systems. All the stages are needed to conserve the processes of life, reconnect people with nature, sustain both nature and civilization, and provide a sustainable supply of resources.

Thoreau's approach to nature can serve us well if we apply it to modern forestry issues. It can help us raise our confusion about forests out of an intellectualized mire and help us consider different ways to obtain forest products. If we adopt Thoreau's approach, we can have forests that are "natural" in the way he accepted wilderness—forests that are self-sustaining, that continue necessary processes of change and succession. His approach offers a path that can lead to sustained production of the forest products people need and desire. It can help us avoid an empty search for a hypothetical, preconceived "expectation" about what is right. If we are honest with ourselves about what we value and what we know, we can have wildness that is truly meaningful to the human spirit *and* improve our use of the tangible aspects of forested nature. In short, Thoreau's sensitivities, understanding, and approach to nature-knowledge can help us engage in practical and environmentally, spiritually valuable forestry.

Chapter 13

Baxter and His Park

"Man is born to die. His works are short lived. Buildings crumble, monuments decay, wealth vanishes but Katahdin in all its glory forever shall remain the mountain of the people of Maine."

Percival Baxter, governor of Maine

As HE WAS FINISHING the journal that documented his second trip to the Maine woods in September 1853, Henry David Thoreau proposed the establishment of a national park. "The kings of England formerly had their forests 'to hold the king's game,' for sport or food, sometimes destroying villages to create or extend them; and I think that they were impelled by a true instinct," he wrote. He went on, "Why should not us, who have renounced the king's authority, have our national preserves, where no villages need be destroyed, in which the bear and panther, and some even of the hunter race, may still exist, and not be 'civilized off the face of the Earth.'" These preserves, he proposed, would be "for inspiration and our own true recreation." If we interpret Thoreau's idea of including a place for "the hunter race" to mean reserving a place for preindustrial cultural practices, we may assume he connected the "instinct" to create a park with the importance of nature for human creativity, recreation, and cultural heritage. Otherwise, we might "grub them all up, poaching on our own national domains." His idea of the purpose of a national park seems consistent with what he had written elsewhere: "The poet must, from time to time, travel the logger's path and the Indian's trail, to drink at some new and more bracing foundation of the Muses, far in the recesses of the wilderness."

Thoreau was among the first to suggest the creation of a national park. As historian Alfred Runte points out, national parks are an American idea. But two of Thoreau's goals for them—creative inspiration and cultural heritage—differed from the reasons that led to creation of the National Park System in the United States. In his book *National Parks,* Runte notes that the earliest equivalents of national parks were royal hunting preserves and military training grounds, which can be traced back to Assyria in 700 B.C.E. The word *parc* in Middle English meant "an enclosed piece of ground stocked with beasts of the chase," held by the king. Parks for the *public* are a modern idea, beginning with urban parks in France in the eighteenth century and spreading to the United States with such famous ones as Central Park in New York City, designed by Frederick Law Olmsted. Work on that park began in 1857, the year of Thoreau's final visit to Cape Cod.

Runte suggests that in the United States, the original motivation for creating national parks had little to do with environmental preservation as we think of it today. He suggests that Yosemite and Yellowstone National Parks were the nation's answer to the artistic and architectural creations of Europe and were part of the development of a national identity. Early advocates of national parks valued the grandeur and beauty of their natural architectural features. They sought to preserve them primarily for their geological formations, although there were certain biological exceptions, especially California's huge, sculpture-like, ancient coast redwoods and the giant sequoias of the Sierra Nevada. The rationale was scenic. The commercialization of Niagara Falls, which decreased the beauty of the scenery, helped promote the establishment of the first national park, Yellowstone, in 1872, a decade after Thoreau's death.

Interestingly, national parks are products of the same Western civilization that is blamed for the destruction of nature. Perhaps these two are part of the same process. Our technology has insulated us from the cold, darkness, and fear of starvation that characterized pretechnological societies—the blackness and the monster from *Beowulf,* Grendel, outside the hearth—and has separated our lives from nature. During the machine age, when this separation was reaching the consciousness of people such as Thoreau, the idea arose to create preserves for what once had been the natural environments within which people lived. As long as one lived within and viewed oneself as part of nature, a national park was not necessary. More than that, it was not conceivable.

Although the national park idea may have been initiated primarily to preserve natural monuments as art and scenery, this may not have been the motivation at the state level. Even before Thoreau went to the Maine woods, the state of Maine had set aside areas for public benefit, beginning in the eighteenth century and continuing into the nineteenth. Most were small areas scattered throughout the state. In the latter part of the twentieth century, the state combined these into a single Public Reserved Lands system covering about 450,000 acres. Of this, about 300,000 acres were consolidated into twenty "management units" ranging from 3,000 to 30,000 acres in eight counties in the west, north, and east of the state.

The present guiding principle of the Public Reserved Lands system states that "full and free public access . . . to the extent permitted by law, together with the reasonable use thereof, shall be the privilege of every citizen of Maine . . . restrictions may be imposed . . . where they relate to the protection of the public health, welfare or safety, or to the protection of the economic interests or natural resources of the state." The state's policy is to provide public access and to "facilitate passage of public vehicular traffic across adjoining private lands." The lands are managed "according to principles of multiple use and sustained yield." These lands appear, therefore, to be held primarily for recreation and for production of timber rather than biological conservation, spiritual inspiration, or preservation of religious traditions, cultural heritage, or scenic beauty.

The development of Baxter State Park, which includes Mount Katahdin, illustrates the confusions, dilemmas, and mixture of motivations regarding humanity and nature in which our society has been immersed. The park is the result of the dedication of a single man. In 1931, Percival P. Baxter, the former governor of Maine, gave 6,000 acres, including most of Mount Katahdin, to the state, establishing a park that included the mountain. Over the next four decades, he continued to purchase land and present it to the state to add to the park. Baxter eventually purchased 200,000 acres, acquiring the last parcel in 1962, when he was eighty-seven, and he left a trust fund to support the care of the park.

Although Baxter left an incredible personal legacy to the state of Maine and therefore to the United States, he was not the first to put forward the idea that Mount Katahdin should be set aside and maintained by the state or federal government. The Indians believed that the mountain was a sacred place, to be feared as well as revered. Katahdin was first described by

Europeans in 1760 when a British officer sighted the mountain from Moosehead Lake. It was first climbed by people of European descent in 1804, so it was truly a newly found mountain at the time of Thoreau's birth.

The few who climbed Mount Katahdin in the first half of the nineteenth century found it as frightening and overpowering as Thoreau did. William C. Larrabee was one of a party that climbed the mountain in 1838. On the descent, he walked along the edge of a ravine at least "a thousand feet deep, and walled up by perpendicular precipices," he wrote. He added that "the scene was intensely sublime," using the terminology of the Romantic poets that meant the scene evoked a feeling of the power of nature and perhaps the power of God, creating an appreciation of the wonders of the Earth's features and therefore of its creations and creator. Wintry winds whirling the fast-falling snow into "many a fantastic drift" made "the blood run chill and the teeth chatter," he wrote.

Recreational use of the mountain had been proposed long before Baxter began his efforts. As early as 1861, a year before Thoreau's death, a man named C. H. Hitchcock climbed the mountain and suggested that a foot or bridge trail be built to the summit, even though by this time only a few people had climbed the mountain. Thoreau, however, did not propose that Katahdin become a park. His experiences on the mountain did not lead him to believe it was a place for him to revisit or for others to visit. It was too primal, not meant for human beings. Although Katahdin greatly affected him spiritually, he did not view it as a place for recreation or as a source of creative inspiration.

Baxter began his efforts to purchase Mount Katahdin for use as a park when he served in the state legislature, in 1918. At the time, a federal Forest Service survey of 154,000 acres in the area of Katahdin found that 12 percent of the land was "virgin growth" that was being cut, 53 percent was burnt-over land, 15 percent was covered with bare rock or stunted trees, and 5 percent contained timber less than 100 years old. Forested nature as *structure* was disappearing. Machine-age humanity sought to retain remnants of it.

Baxter deeded the park to the state in many separate parcels over the four decades it took him to purchase the lands. Over time, his deeds of transmission to the state gave differing limitations on the use of the parcels and different reasons for his establishment of the park. From these docu-

ments, it appears that his ideas about the purpose of the park seemed to vary over the decades.

With his first gift, in 1931, Baxter stipulated that the land "shall forever be left in the natural wild state, shall forever be kept as a sanctuary for wild beasts and birds, and . . . no roads or ways for motor vehicles shall hereafter ever be constructed thereon or therein." This very modern phraseology seems to suggest some intrinsic value in that landscape—a moral position, perhaps—as well as aesthetic and ecological concerns, to put the statement in present-day terms. Spiritual and inspirational effects of the mountain may have motivated Baxter, but he did not say. According to John W. Hakola, who wrote *Legacy of a Lifetime: The Story of Baxter State Park,* "no single issue concerning the Park has created so much controversy for so long a time as that of an adequate definition of [Baxter's term] 'natural wild state.'"

With subsequent gifts of land, contingencies imposed by Baxter focused on recreational use (other than hunting) and protection of wildlife. At one time, he purchased burnt-over lands to be used both for recreation and as future sources of timber. Some of his later statements allowed for construction of roads that would lead to camps and cabins. He varied in his decisions about access for motor vehicles and the extent to which the park was to preserve nature or be available for recreational use.

Although Baxter vacillated in establishing permissible uses of the park, he understood the wilderness idea as it is used in today's context, as indicated by a statement he made in 1941: "Katahdin always should and must remain the wild stormswept, untouched-by-man region it now is: that is its great charm. Only small cabins for mountain climbers and those who love the wilderness should be allowed there, only trails for those who travel on foot or horseback, a place where nature rules and where the creatures of the forest hold undisputed dominion." In 1945, however, he made clear that his use of the phrases "natural wild state" and "sanctuary for wild beasts and birds" did not mean that the park would be "a region unvisited and neglected by man." He was trying to provide a place without commercial exploitation, hunting and killing, lumbering, hotels, advertising, "hot-dog stands, motor vehicles," and "the trappings of unpleasant civilization," but he wanted the park to be accessible to the public. Moreover, he made clear that the park was not to be only for the elite: "Nor is the Park to be kept exclusively for professional mountain climbers; it is for everybody," he said. He wanted the park "left simple and natural" and "as nearly as possible as it was when only the Indians and the animals roamed at will through these areas," and he wanted it available to "persons of moderate means." He wrote, "I do not want it locked up and made inaccessible; I want it used to the fullest extent but in the right unspoiled way." He was, of course, aware of the development of national parks elsewhere in the United States, but he resisted the suggestion that Katahdin be one, arguing that the state of Maine should be able to do at least as good a job with its own natural heritage as would the federal government.

It is not surprising that Baxter's ideas about the park changed over time or seemed full of ambiguities and contradictions. After all, it was not clear at that time what was necessary to maintain a park in its natural state—nor is it completely clear today. At first, the goal seemed to require no more than putting land aside and preventing logging, hunting, and motor vehicle traffic. Later, though, Baxter began to see the problems that arose when land was otherwise untouched by human managers. When fires were prevented, the forests changed; if a fire did occur, the land then seemed in need of active restoration to promote reforestation. These and other effects suggested that the "natural state" might require some human restorative actions, which would seem to contradict his earlier statements. He was trapped by a lack of three essential elements: scientific information, a clear concept of the relationship between human beings and nature, and an understanding of natural ecological processes—the dynamics Thoreau found on Cape Cod and in the woods near Concord. Lacking better scientific understanding, Baxter had to muddle through with the best advice and knowledge he could obtain and his own best guesses.

However, there seems to have been a central thrust to his goals: to create a park that *appeared* to be natural and *was* in fact as natural as possible, whatever that meant—but that allowed public access. But how could one create a "wild stormswept, untouched-by-man region" that would forever be "a sanctuary for wild beasts and birds" yet allow public access?

Baxter referred to wildness as the "great charm" of the mountain, which is to say that it was the human experience of the mountain that was key, not wildness per se but its charm as scenery. *Charm* is a limited word, much different from *sublime* or *awe-inspiring*. The latter terms might suggest a place that had "fantastic drifts" of snow that make "the blood run chill and the teeth chatter," as described by Larrabee in his 1838 ascent. *Charming* might describe a homespun tapestry of Mount Katahdin hung in a kitchen, not the actual experience of snow, wind, clouds, and precipices. Did Baxter consider charm a quality consistent with a mountain that is a "sanctuary for wild beasts"? And how could he have believed that the mountain could be "untouched by man" even as it provided cabins and trails for those who wanted to ascend it?

Perhaps Baxter would have been able to clarify his thinking if he had had the opportunity to speak with Thoreau or, better, travel with him through the Maine woods. The park is large enough to allow for a variety of uses in different areas. Baxter seemed to have realized this when, late in life, he gave the state a portion of land in the northwestern part of the park for demonstration of modern silvicultural techniques—scientific methods of forestry. He wrote that this area "shall forever be held for and as a State Forest, Public Park and Public Recreational Purposes and for the practice of Scientific Forestry, reforestation and the production of forest wood products." All would be done "according to the most improved practices of Scientific Forestry." The revenue would be used to maintain the park. Given the total area of Baxter State Park, his other goals could have been met in it as well.

I have been asked: Why does it matter that Baxter changed his mind about the purposes of the park and that the deeds to his twenty-eight gifts to the state, written over a period of thirty-two years, are worded in various ways? Is it not sufficient that the land was put aside, whatever the rationale? With the land secured, one might imagine, nature could take its course, and the preserve would be safe and sound. From this perspective, any means—any wording of deeds—would justify the ends: the creation of nature preserves and parks. However, the present director of the park, Buzz Caverly, finds that the wording of deeds does matter. He said that his job "constantly focuses" on this concern. He deals with continuing conflicts over current desires and Baxter's original wording. Indians have appealed to the Baxter State Authority to obtain some of its lands for their exclusive use, as have the Boy Scouts of America. "While I sympathize with their desires, I cannot allow these exclusive uses to happen because they violate Baxter's deeds," Caverly said. An organization called the International Appalachian Trail/Sentier International des Appalaches (IAT/SIA) would like to have the right to make decisions about the famous trail where it passes through the park, but this, too, would violate terms in Baxter's deeds. The IAT/SIA would use its decision-making power to promote nature conservation and assist with recreation in the park, goals that seem consistent with those of the park. This conflict in particular shows how much the wording matters. The Americans with Disabilities Act creates yet another conflict for park administrators because it requires certain kinds of access for disabled people, calling for structural modifications that also violate the terms of the deeds. Thus, in both day-to-day management activities and the larger issues that become lawsuits, the importance of Baxter's wording, and of his changes in wording, is evident.

It is interesting that I would be asked about the importance of Baxter's wording regarding the conservation of nature. I have tried to imagine other situations in which confusion about means and goals would be acceptable. Suppose, for example, that you boarded an airplane and heard the pilot announce over the intercom that he was somewhat confused about where the plane should go and how to fly it. Would you stay on that plane? Or suppose you consulted a physician and he confessed to feeling confused that day and was uncertain what to do for you—would you find that satisfying? If an architect or a landscape architect brought drawings for you to review but then expressed confusion about the design of your house or garden, or brought quite different and inconsistent plans at different times, would you be comfortable in having him continue designing your home or landscape? Of course not. Yet when it comes to nature and wilderness, we are so deeply affected by the idea that "nature knows best"—and the assumption that we are so separated from nature that *we* could not know what is best—that we allow ourselves to believe that our confusion is acceptable: nature will always bail us out.

Because of the large size of the park and the variety of wording in the many deeds, a number of options exist for designating uses in the park. There is ample room for roadless areas, open only to motorless travel; there is room to restrict even these uses to conserve certain endangered species or ecosystems; and there is room, as explicitly stated by Baxter, for an area in which to demonstrate the best methods of silviculture. The optimal layout for the park is a landscape architectural design problem, not a matter of pursuing a single truth. All justifications for preserving natural areas could be met within the park, some in distinct areas.

After his travel to Mount Katahdin, Thoreau rejected its barren summit as an inspirational place in the constructive sense of promoting creativity in the arts. But the ascent was an important spiritual experience for him nonetheless. The physical attributes of the mountain affected his perception of the intangible, religious, and spiritual aspects of human experience. Katahdin drove Thoreau away from wilderness per se as a place for inspiration and creativity and toward the partially settled countryside. It was there, in his walks near Concord, that Thoreau found creative and inspirational contact with nature, in a piece of wilderness, a swamp on the edge of town—what we today would call a suburban park. Today, conquered and experienced by many, Mount Katahdin is more a place of physical recreation than a location for spiritual inspiration—although the mountain doubtless has a spiritual effect on many who reach its summit and experience its wind, clouds, storms, boulders, and barrenness.

The story of Baxter State Park seems to be the best case one could imagine. Baxter—a man of goodwill who wished to create a park for public benefit and to conserve nature, a lawyer and governor who understood legal wording and its implications—stands out as a hero for creating a park for people and for nature in a place of great beauty. Baxter recognized the potential future problems that could accompany changes in social goals and new kinds of pressures on the land. He created a park that is much used and appreciated. In this sense, he succeeded. But he was not able to avoid contention, nor did he appear to succeed in all his goals. The question the story raises is, If Baxter could not succeed completely, who could? I suggest that Baxter could have profited from Thoreau's help in clarifying his questions and approaching the design of a landscape that would meet the needs of both nature and people.

The history of Baxter and his park is much more than a discussion of an environmental and political issue. It reaches to the depths of human existence and ideas. We create parks as much to save ourselves as to save nature. It is clear that wording and intention do matter; that confusion in means and goals does allow conflicts and create difficulties in achieving the conservation of nature and providing public benefits of nature. If we confuse the purposes of parks and therefore confuse both our desires and the execution of our plans and actions, we also confuse ourselves. Baxter's story suggests that we today, dealing with our deep concerns about the relationships between civilization and nature and the individual and nature, can profit from Thoreau's approach. His approach to understanding nature can help us with the preservation of natural areas as much as it helps us with the utilization of natural resources.

Chapter 14

Creating Wilderness

"Just before night we saw a musquash. . . . The Indian, wishing to get one to eat, hushed us, saying, 'Stop, me call 'em;' and, sitting flat on the bank, he began to make a curious squeaking, wiry sound with his lips, exerting himself considerably. I was greatly surprised,—thought that I had at last got into the wilderness, and that he was a wild man indeed, to be talking to a musquash! I did not know which of the two was the strangest to me. He seemed suddenly to have quite forsaken humanity, and gone over to the musquash side."

Henry David Thoreau, *The Maine Woods*

HENRY DAVID THOREAU'S ideas about wilderness leave us with a conundrum. A 1964 federal law, the Wilderness Act, defines wilderness as a place "untrammeled" by human beings, where people are only visitors. The act states that a wilderness area "generally appears to have been affected primarily by the forces of nature, with the imprint of man's work substantially unnoticeable; . . . has outstanding opportunities for solitude or a primitive and unconfined type of recreation; [and] has at least five thousand acres of land or is of sufficient size as to make practicable its preservation and use in an unimpaired condition." Aldo Leopold wrote of wilderness as a place where one could walk for two weeks without retracing one's own footsteps. The Boundary Waters Canoe Area Wilderness, mentioned specifically in the Wilderness Act, contains about one million acres. In general, a wilderness area in the United States is conceived to be large, remote, and separated from the imprint of humanity and civilization.

Yet Thoreau, having experienced the vast wilderness in Maine, rejected large areas of wilderness. The wildest, rawest wilderness he visited, the summit of Mount Katahdin, had a great effect on him at the time, leading him to reject that kind of wilderness experience because it did not provide the spiritual and creative contact with nature he sought. Even though his climb up the mountain greatly affected him at the time, Katahdin does not appear in his printed works as a major image or symbol. In fact, Thoreau rarely mentions Mount Katahdin elsewhere at all, although mountains do appear as images and symbols in some of his works.

In contrast, it was the biologically rich swamp, surrounding him so closely with life that distant vistas were obscured and size became irrelevant, that held the deepest meaning for Thoreau. The swamp, with its deep soil made of molding vegetation, reappears throughout his writings as a major image and symbol. It was the richness and bountifulness of life, not the barren rocks and piercing winds of an unvegetated outcrop, that inspired Thoreau.

The important idea about wilderness that we obtain from Thoreau is that a person can experience wildness, with all its spiritual, religious, inspirational, and creative benefits, even in a small area in which the effects of human actions are quite apparent. This idea is quite different from our modern view.

The conundrum is: How can we reconcile our modern idea of wilderness with the continual division of landscapes as the human population grows and the effects of technology touch every piece of the Earth? In the United States, people tend to believe that bigger is better and the country is what it is because it has the biggest and therefore the best—the biggest canyons, the biggest trees, at one time the tallest buildings, the longest bridges. Have we come to the same belief about wilderness—that bigger must be better and only a big wilderness can provide an experience of the wild? Thoreau was able to find solitude on a walk a few miles from Concord and in a swamp on the edge of town. A small wildness provides benefits. It allows greater access for more people to the benefits of contact with nature. It can be brought to the people; it might even be within or near a city. People and nature can be more easily integrated.

But was Thoreau's rejection of vast, unpeopled wilderness in some way a limitation of his imagination rather than a profound assessment of the true character of wilderness?

Late in his life, Thoreau seemed to vacillate on the question of how large a natural area should be. During his last decade, when he made daily walks of about four hours, revisiting the same tree, shrub, or flower to make scientific measurements, he wrote an essay called "Walking." This essay contains some of his last, and some might suggest most mature, thoughts about human beings, civilization, and nature.

The presence of some *uncultivated* land was essential to Thoreau. In "Walking," he wrote, "I would not have every man nor every part of a man cultivated, any more than I would have every acre of earth cultivated; part will be tillage, but the greater part will be meadow and forest, not only serving an immediate use, but preparing a mould against a distant future, by the annual decay of the vegetation which it supports." But meadow and forest, where Thoreau found wildness, were often heavily modified by human action, and therefore would not have met the requirements of the Wilderness Act.

In the same essay, Thoreau wrote, "In short, all good things are wild and free." But Thoreau could receive the *experience* of wildness in those daily walks near to home. "When I would recreate myself," he wrote—and here I take *recreate* to mean both recreation and re-creation—"I seek the darkest wood, the thickest and most interminable and, to the citizen, most dismal, swamp. I enter a swamp as a sacred place, a sanctum sanctorum. There is the strength, the marrow, of Nature. The wildwood covers the virgin mould, and the same soil is good for men and for trees. A man's health requires as many acres of meadow to his prospect as his farm does loads of muck." But this swamp need not be large nor far from civilization; Thoreau did not require the vastness Leopold described. Having experienced the Maine woods in three trips, he concluded that he preferred Concord and its surroundings, "the partially settled countryside" that he considered best. In his journal entries about his daily walks during the last ten years of his life, Thoreau often referred to "wilderness" that contained people, houses, and other obvious results of human activities, and he wrote of finding contact with wildness on these walks.

My own reactions to wilderness have been largely similar to Thoreau's but in some ways different. I have found satisfaction in small areas of wildness affected by human actions, but I have also deeply appreciated a large wilderness free of almost all human influence. Over a five-year period, I conducted research on moose and their ecosystems at Isle Royale National Park with a colleague and friend, Peter Jordan, and I found a kind of magnetism in that wilderness. On each visit, after a few days' adjustment, I found the wilderness fascinating. Each tree in the forest, each branch on the ground, each fox, moose, or squirrel told a story that one might unravel. Why were they at that spot at that time? How did they get there, and what had led to their current condition? The wilderness was at once a mystery and the source of a deep feeling of contact with nature. When I left the island, I often experienced "withdrawal symptoms," wanting to turn around and go back immediately rather than return to civilized life.

No doubt there are many human responses to wilderness, both within and among individuals, ranging from sheer terror to serenity. Another friend and colleague, Lee Talbot, one of the acknowledged world experts on wildlife and wilderness, conducted some of the earliest modern research on Africa's Serengeti Plain. He has worked on environmental issues in more than 120 countries and spends some time each summer with his wife, Marty, hiking through large stretches of wilderness. For Lee, vastness is an essential quality of wilderness, an important part of his life. Like Thoreau, he does not choose to live within that wilderness. But unlike Thoreau, he feels a need for contact with an area large enough that he can feel completely removed from civilization, for at least a short time each year.

But few people seek such vast wilderness. In forty years of fieldwork at Isle Royale, Peter Jordan has never encountered a visitor more than a hundred yards off a trail. Most visitors to legally designated wilderness areas stay on the trails or the well-defined waterways, making their camps—as is required in the Boundary Waters Canoe Area Wilderness—in designated and therefore human-altered campsites.

Recently, I visited the Plum Creek Timber Company in Maine, where forester Carl Haag took me to a plantation of mature spruce. The trees were evenly spaced, and the area between them was open because of the dense shade they created and the deep layer of needles on the ground. It was quite pretty. Carl told me that the company had recently conducted tours for the public on its lands, including this plantation. One woman taking a tour had insisted that the area was natural and could never have been a plantation, regardless of corporate records documenting its status as a former farm and indicating when trees had been planted, and despite the fact that natural dispersal of seeds could not have resulted in trees so evenly spaced. It was the beauty of the stand that persuaded this visitor that it was natural. This woman did not need wilderness in the way the Wilderness Act defines it: for her, a beautiful plantation, heavily "trammeled" by people, was nature enough.

Perhaps one could speak of a complete lifetime experience of wilderness. This would involve some contact with vast, largely untrammeled areas—just like Thoreau's trips to the Maine woods. But it would consist primarily of experiences in "small wildnesses"—areas nearby, conveniently reached, where one would find most often the inner meaning of nature. These could be natural or naturalistic, free from human influence or affected by human actions to some extent. The kind of human actions affecting the "small wildness" must be consistent with the mood the area creates. In the woods near Concord, for example, Thoreau saw a particularly pretty cabin to be part of the beauty of the landscape yet found another building ugly. This resolves the conundrum: the complete person would appreciate vast wilderness but also be capable of receiving a spiritual, uplifting experience of nature in the altered woods next door.

An alternative resolution of the conundrum would be that the establishment of vast wilderness areas would become a major societal goal, though understood to be a luxury available only to a prosperous and open society intent on providing multiple benefits to its citizens. The creation of such areas, however, carries other burdens. A society cannot maintain very large wilderness areas and simultaneously allow unlimited human population growth.

Thoreau wrote: "Life consists with wildness. The most alive is the wildest. Not yet subdued to man, its presence refreshes him." If one confuses wildness with wilderness and wilderness with vastness, then one might misinterpret this assertion to mean that only large areas of wilderness are valuable. But Thoreau clearly distinguished between wildness and wilderness and made no reference to the size of an area that would be of value to him. Once again, for Thoreau, land that was "wild and free" included aspects of civilized life that we do not associate with wilderness today. Continuing on the same theme but emphasizing that he could obtain a sense of wildness in a partially settled countryside, Thoreau noted in the same passages, "There is something in a strain of music, whether produced by an instrument or by the human voice,—take the sound of a bugle in a summer night, for instance,—which by its wildness, to speak with satire, reminds me of the cries emitted by wild beasts in their native forests." This passage harkens back to his experiences as a young man, canoeing on the Concord and Merrimack Rivers, when at one of his night camps, the sound of a dog barking was to him the essence of wilderness.

For Thoreau, the *experience* of wildness was the key. "In literature it is only the wild that attracts us," he wrote. "Dullness is but another name for tameness. It is the uncivilized free and wild thinking in Hamlet and the Iliad, in all the scriptures and mythologies, not learned in the schools, that delights us. As the wild duck is more swift and beautiful than the tame, so is the wild—the mallard—thought, which 'mid falling dews wings its way above the fens.'" But again, for Thoreau, wildness could be a view of an attractive cabin by the Concord River near sunset, in a dense but probably second-growth forest.

Chapter 15

Conserving Wilderness

"I wish to speak a word for Nature, for absolute freedom and wildness, as contrasted with a freedom and culture merely civil,—to regard man as an inhabitant or a part and parcel of Nature, rather than a member of society. I wish to make an extreme statement, if so I may make an emphatic one, for there are enough champions of civilization."

Henry David Thoreau, "Walking"

THE ISSUE of the optimal size and shape for a national park has come home to the Maine woods in which Henry David Thoreau traveled. Recently, an organization called "RESTORE: The North Woods" proposed the establishment of a Maine Woods National Park and Preserve, which would cover 3.2 million acres in north-central Maine and cost an estimated $500–$900 million to purchase. The park would include areas of considerable fame and recreational appeal: the headwaters of Maine's major rivers; half of the Allagash Wilderness Waterway; most of the Penobscot River, under study for inclusion in the National Wild and Scenic Rivers System; and 100 miles of the Appalachian National Scenic Trail. Baxter State Park, however, would not—could not because of Percival Baxter's statements in his deeds of gift—be included within this national park.

Is this big enough, too big, or just right? The intangible values of wilderness might help us decide. Let us consider Aldo Leopold's idea of a wilderness as a place big enough that one could walk for two weeks without retracing one's footsteps. Walking at a rate of 2 miles per hour—a good pace for a cross-country skier away from trails and carrying a heavy pack or even an individual traveling by horseback or canoe—for ten hours per day would transport a person 20 miles. If we imagine a circular area that would bring our hypothetical traveler back to the starting point in two weeks, the circumference of the circle would be 14 times 20, or 280, miles. The circle's diameter would be 89 miles, and its area would be 6,218 square miles, or more than 3.98 million acres.

For comparison, Isle Royale National Park on Lake Superior, which meets the conditions of an untrammeled place where people are only visitors, contains approximately 280 square miles, or 179,200 acres. The Boundary Waters Canoe Area Wilderness, one of the most famous (and one of the first) wilderness areas designated under the Wilderness Act, covers 1,563 square miles, or approximately 1 million acres—one-fourth of the circular area that would meet Leopold's suggestion. Of course, one could walk back and forth without retracing one's steps in smaller areas, such as a series of heavily wooded parallel ridges and valleys, in which one could follow the valleys or ridges. A landscape architect could design a wilderness that would take only one-third to one-fourth of a circular area—1 million to 1.3 million acres. This is about the size of the Boundary Waters Canoe Area Wilderness.

It would be convenient if scientific information gave us a clear idea of what is necessary to sustain ecological systems and enhance the probability of the persistence of all life on Earth. Typically, justifications for establishment of wilderness areas, nature preserves, and new parks seek such a foundation. This is true for the proposed Maine Woods National Park and Preserve.

The park, with 3.2 million acres, would be bigger than the state of Connecticut and the "greatest nature reserve east of the Rocky Mountains." Land for the park would be purchased only from willing sellers, and purchases would concentrate on "a few large paper and timber company holdings, which make up most of the proposed park area." The park would be large enough to "include entire ecosystems." This last statement suggests that entire ecosystems are very large, but that is not correct: ecosystems vary greatly in size. For many years, ecologists have studied areas smaller than 100 acres that are considered to be ecosystems.

The problem with searching for a scientific basis for the necessity of wilderness as defined in the Wilderness Act is that ecology is still a young science that provides little specific guidance. What is required to sustain any species or set of species is poorly understood.

On a recent sunny September day, I went to Big Reed Pond Preserve in Maine with Mac Hunter, professor of wildlife at the University of Maine, who kindly acted as my guide. The preserve covers about 5,000 acres and includes woodland along the pond. It is the only large tract in Maine known never to have been logged. In terms of land area, the preserve meets the Wilderness Act's definition of wilderness.

There are no trails in Big Reed Pond Preserve. Mac and I hiked cross-country from an old logging road down to the pond. We followed compass directions more or less, but we mainly followed Mac's sense of which way to go based on previous trips. We were in no hurry and stopped frequently to examine the woods and the soil to see what this uncut forest was like. We found that it had some large but not huge trees, and many larger ones had blown down. Mac said that there had been a storm a number of years ago; perhaps it had brought down the older trees. But on further observation, we saw that the downed logs lay in many directions, so the cause did not seem to have been a single storm.

Storms are a natural event in these woods, a way in which forests lose standing but unhealthy trees. Closely examining the downed trees, we noted that some had split off about waist high, which meant that the roots were strong and the wind had been sudden and forceful, acting like a knife to cut the stem from the base. These trees were probably quite healthy when the storm felled them. Other trees had fallen with their mass of roots still attached at the base of the stem. The "root bowls" lay exposed along with the soil that had clung to the main roots, leaving a small pit that had been excavated as the tree toppled over. These trees had been unhealthy, with dying roots.

The roughness of the undulating terrain told us that no plow had cut the land. We saw no sign of logging. But the forest lacked the idealized look of old growth pictured in many books and on television—vast areas of huge, widely spaced trees.

We ate our lunch at the edge of the pond, which was hard to see because of low branches of large cedar trees that extended from shore. We found a few spots near the bases of the trees where we could peek out at the water. It was a pleasant scene. Across the pond, a small float plane was parked, but there was no a sign of any other person. We saw no one else and heard no other people.

On our way back, we heard a noise ahead of us and walked quietly, looking for the source. We soon saw a large black bear about twenty or thirty feet up in a beech tree, feeding on beechnuts. Mac said that in all his years of studying wildlife in Maine, he had never seen a large bear up a tree like this. We slowly approached, and the bear finally noticed us, came down the tree, and went off in another direction. Although we saw the bear near the edge of the preserve, we believed that this was a fortunate occurrence, that the bear probably moved in and out of the preserve. Sighting the bear was the most remarkable experience of our walk. Overall, it had been a pleasant experience but not especially inspirational, and the scenery had not been structurally grand.

What conditions would be necessary for Big Reed Pond Preserve to sustain itself as it appears today? First, the 5,000-acre preserve would have to produce enough seeds to keep up with storm damage: given the high proportion of wind-toppled trees, seed production would have to be sufficient and habitats adequate for the seeds to germinate and grow. Over the long term, addition of young trees would have to be sufficient to replace mature ones that die. Further, water levels would need to remain within a range that allows trees to persist both in the wetlands and on the uplands. The twenty-four chemical elements required by living things would have to be available at the right times and in the right amounts and proportions, and habitats would be needed for the soil-dwelling fungi and bacteria that help trees take up chemical elements and decompose organic matter. Finally, for bears to persist, we would have to estimate the minimum population size of bears that could sustain itself and the size of an area required to provide their food and shelter.

At present, however, there is not a sound scientific basis for determining the minimum land area needed for a particular type of ecosystem to sustain itself. As a result, there is a tendency to play loosely with the ecosystem concept, as seen in the proposal for Maine Woods National Park and Preserve. People end up dealing with neither actual size nor number of ecosystems, instead appearing to argue that only an area of millions of acres would be sufficient—an argument for which there is not adequate scientific justification. This is inconsistent with the scientific method, in which concepts must be clearly defined and understood, and quantities measured.

A second scientific justification put forward for establishing this park is that "many top-predators native to the Maine Woods such as the extirpated wolf and imperiled lynx, require large areas of wildland to survive." As a result, the large area of the proposed park and preserve "would insure the permanent protection of essential habitat for these kinds of species." Although bears are omnivores and not necessarily top predators, a large area would need to be protected to sustain bears at Big Reed Pond.

The park proposal states that "large carnivorous predators perform a key function in the health of an ecosystem. They influence the distribution, numbers and age structure of prey species like the moose and deer. To quote the poet Robinson Jeffers, 'What but the wolf's tooth whittled so fine the fleet limbs of the antelope?' The protection of large predators would further ecosystem recovery in the North Woods." The idea that top predators precisely regulate the balance of prey populations in an ecosystem is, however, a myth. There is so little solid scientific justification for it that the writer is reduced to citing a poet as a supposedly scientific expert.

If predators such as wolves are necessary in the North Woods, how have moose and deer survived since the wolf was locally extinguished? Following their introduction into new habitats, with or without the presence of predators, large ungulates such as moose and deer undergo a consistent pattern of change in abundance, rapidly increasing to a large population size and then crashing to a much lower one. Afterward, population size varies over time but never equals the first peak.

Suppose we grant that top predators *are* necessary in the proposed park. How large an area do they need? The scientific basis for determining the area required by wolves is weak. People eliminated wolves from most of their original habitat before the advent of modern science or before scientists could get to the wolves to study them. There are two questions here: what is the average area required by a wolf, and how many wolves form a minimum population that has a low risk of extinction and is likely to persist—a minimum viable population?

A few examples illustrate the range of areas used by large mammalian top predators. I present these not as a tight scientific argument but as an illustration of the range of possibilities known at this time. In the late 1940s, a pack of wolves migrated to Isle Royale National Park, and the population of the pack varied from 10 to 30. For some years, the wolves declined in number because of a viral disease, probably transmitted by domestic dogs illegally brought to the park by visitors. But today, the wolf population has recovered and appears to be sustaining itself. Given that the island covers 280 square miles, a population of 30 wolves means that the average area per wolf is 9 square miles. A population of 10 means 25 square miles per wolf. However, Isle Royale has a large population of moose and a sizable population of beavers. The moose population, varying from 1,500 to 3,000, has the highest density known anywhere in the world, at 0.2 to 0.1 square mile per moose. A wolf's required habitat area depends in part on the population density of its prey. Other factors include behavioral factors, such as the territorial behavior of a pack of wolves, which could involve a minimum area, and the availability of den sites and other habitat elements.

To determine the total area needed to sustain a population of wolves, we need to know the minimum number of wolves that make up a self-sustaining population and the minimum area required per individual. In standard practice over the past quarter century, the number required for a minimum viable population of any species over the long term has been estimated to be 500 individuals. The estimate was originally based on a study of the genetics of the domesticated cow, and it is unclear how well it reflects the requirements of a wild population. The argument for this number is limited to genetic requirements: scientists assume that a population of this size would not suffer from genetic problems that can occur in small populations. Assuming that this estimate is appropriate for wolves, a population of 500 wolves, each requiring 28 square miles, would need 14,000 square miles. Maine occupies 33,265 square miles, so the required habitat would be approximately 42 percent of the land area of the entire state.

Another study reports that the average home range for adult male wolves in Wisconsin is 124 square miles. Assuming that this is the range of a lead male and therefore the range of a pack that might on average number about 10 animals, the area per wolf might be lower—about 12 square miles per wolf. If each wolf required 12 square miles, then only 6,000 square miles would be required, or 15 percent of the land area of Maine. If the required home range is 124 square miles per wolf whatever its social position in the pack, then 62,000 square miles would be needed—more than the entire area of Maine. Clearly, given the large variation in the estimate based on present thinking and weaknesses in current scientific understanding, we do not have a definitive scientific basis for making a decision. We must make a decision with incomplete scientific data and understanding or reasons other than scientific ones.

The argument for basing the size of a preserve on the number of wolves rests on four assumptions. First, the presence of wolves precludes other land uses such as logging and any human settlements. Second, the population of wolves in the state of Maine would be self-contained, with no reliance on populations in adjacent Canada or Vermont. Third, the only way to maintain a wolf population is to leave it completely alone. And fourth, the area required is independent of the density of food supply. In reality, the higher the food population density, the less hunting area a wolf would need to find enough food. The required home range can be expected to vary with the population density of prey as well as with other varying environmental factors.

There are alternatives to the park proposal that could sustain a population of wolves. For example, wolves may be able to persist alongside some kinds of human activities. Or, with modern methods of wildlife management, it may be possible to have a much smaller population and occasionally move a few wolves to different locations—from one state or from Canada to Maine and a few from Maine to other places. This could be done very infrequently, perhaps at a rate of ten wolves every ten years. The area and population required to sustain wolves in Maine could be reduced further. My point here is not that a decision is impossible but that we must recognize the importance of nonscientific reasons and learn how to use incomplete scientific data in our decision-making processes.

The lack of adequate scientific information and the misuse of existing information in this instance has an interesting parallel with the work of Wilbur and Orville Wright in developing the first airplane. The parallel is apparent in a fascinating biography of the Wright brothers, *Kill Devil Hill,* by Harry Combs and Martin Caidin. When the brothers first began investigating the possibility of building a flying machine, they were daunted by the long list of great minds throughout history who had attempted to discover the secret of flight, and they questioned their own abilities to make any contribution. But then they began to look for useful data from the centuries of thought. Wilbur soon found that there was little of value. There was no "flying art" in the proper sense of the expression, he wrote, only a "flying problem":

> Thousands of men had thought about flying machines, and a few had even built machines which they called flying machines, but these machines were guilty of almost everything except flying. Thousands of pages had been written on the so-called science of flying, but for the most part the ideas set forth, like the designs for machines, were mere speculations and probably ninety percent was false. . . . Things which seemed reasonable were very often found to be untrue, and things which seemed unreasonable were sometimes true . . . things which we at first supposed to be true were really untrue . . . other things were partly true and partly untrue . . . a few things were really true.

As a result, the Wright brothers had to obtain the necessary empirical data and develop the theory of flight, including the mathematics of airfoils and propellers. In working out their scientific problem, they followed the same approach I have been describing for Thoreau, although they did not have to deal explicitly with the spiritual and inspirational issues—these, expressed as a fascination with flying, were assumed and accepted implicitly. Their experiences should serve as a warning to us today as we approach the much more complex systems that we loosely call wilderness and even more loosely claim to understand. We should remember Will Rogers's admonition: "It ain't what you don't know that gets you, it's what you do know that ain't true." In the desire to justify the preservation of large areas of wilderness by any means, there is a tendency for people to fall victim to the fallacies the Wright brothers found when they investigated the "science" of flying.

The problem with the lack of adequate information extends to justifications based on the economic value of wilderness and natural areas. The proposal for the Maine Woods National Park and Preserve claims that there would be economic benefits, using the analogy that Acadia National Park "brings more than $100 million to the surrounding communities each year." This is a dangerous comparison, however, because Acadia is located on the beautiful Maine coast, has ready transportation access, and is often visited as part of a longer vacation trip. Much of the 3.2 million acres of the proposed park and preserve would be in remote flatlands of much less appeal. Few national parks attract $100 million worth of business per year. The national parks most often visited are the Washington Monument and others near urban areas. Olympic National Park, in Washington State, attracts 5 million visitors per year, and Yellowstone National Park attracts 3 million; they are two of only a handful of national parks that include areas that fit the modern conventional idea of wilderness and also attract visitors in the millions. Thus, the basis for the economic argument is weak.

The proposed park and preserve would consist of areas set aside from human intervention as well as other areas, called "preserves," where people could hunt, trap, snowmobile, and participate in "other traditional recreational uses." There would be no commercial development and therefore no discussion of how residents of the state of Maine would make large amounts of money from the park. It may be that a Maine Woods National Park and Preserve would attract millions of visitors and lead to an economic gain of $100 million, but experience elsewhere suggests that this is unlikely, given the terrain and competition from existing recreational opportunities.

Jym St. Pierre, Maine director of RESTORE: The North Woods, gave further justifications for the park in an article in the organization's newsletter. He wrote about his enjoyment of the Maine woods, lakes, and seashores as a child and his reaction to their loss in favor of roads and "suburban" development: "I have been shocked to fly into favorite remote spots and find the wildness impaled by new roads." And what would replace this, then, in a park supposed to bring in an equivalent of the $100 million of Acadia? Would the public, except individuals who could travel by plane, be excluded? Would hundreds or thousands of visitors be flown into these favorite remote spots—would that preserve the wilderness character? There is an element here of an elitism that does not appear in Thoreau's writings but that has been a factor in the establishment of other parks. The original parks, the hunting preserves of kings, were clearly elitist, but in the United States, a national park is supposed to be for the people. One could argue for the creation of roadless and pathless nature preserves within a park of 3.2 million acres, but this would permit even the airplane visits of a few to their favorite places.

In another article in the RESTORE newsletter, Jon Luoma describes cross-country skiing in the clear-cut forests of the Maine woods as a process of snaking through "mazes of stumps and over heaps of abandoned slash, or [struggling] dispiritedly along muddy skidder tracks and logging roads." He and his skiing companions, he wrote, could find no "real woods." The implication is that errors of the past will be corrected by establishment of a gigantic park. But an intelligent mixture of logging and conservation areas would take care of the cross-country skiing problem. A plan to accommodate both uses requires quantitative analysis, landscape design, and a detailed understanding of the woods.

The articles defending the proposal for a Maine Woods National Park and Preserve present a confusion of justifications based on a weak foundation of facts. Value judgments are clearly embedded in the proposal: that the value of wolves and their ecosystems exceeds the value of all other land uses; that only certain kinds of recreation are compatible with the maintenance of wolves; and that in a democracy, the majority of people will favor such a change in land use. This is not the approach that Thoreau used, either in seeking facts or in discovering intangible values.

The problem with the park proposal is not in the attempt to use science as a justification; it is in the improper use of science. We must use science to determine the area and shape of preserves that meet the needs of endangered species, protect biological diversity, and permit each particular geographic portion of life to play its proper role in the Earth's life-supporting and life-sustaining system. But we must act from knowledge in the way Thoreau sought to do: through the proper use of science and expertise.

Underlying the arguments for creating the park and preserve I have detected a belief that the scientific arguments will be taken seriously, whereas spiritual, creative, uplifting, and recreational justifications will not. The tendency is to seek scientific justifications even if they are weak. Given this situation, we need to act in accordance with the precautionary principle—the less we know, the more cautious we must be in our actions, especially where the results of these actions are likely to be irreversible. A conservative approach is to save large areas until the necessary sizes can be determined. But this forces a lot on a still emerging science. Because of weak but desired scientific evidence and the mixture of tangible and intangible motivations for preserving natural areas, justifications for doing so tend to take on a puritanical tone: that nature is good and people have been evil, that people must be kept from the sin of destroying nature, and kept separated from it.

The two approaches to determining a suitable size for a wilderness—using scientific information and applying intangible values—leave us in an uncomfortable position. At this time, neither science nor intangible values can tell us what such a size *must* be. But right now, until we obtain more scientific information, the intangibles combined with the precautionary principle are likely to work better. Here, the methodology I have described as Thoreau's becomes important. We must seek to know the facts as they are, not with "expectation." We must integrate the tangible and intangible values. We must listen to experiential and professional experts to gain a starting point for our consideration. We must look deep within ourselves for what is truly valuable.

This discussion of the problems inherent in determining the appropriate size of wilderness areas is not meant to be an excuse to deny the need to establish future wilderness areas or to withdraw from present proposals. Rather, it is intended to suggest that we should not be so distracted by our quest to set aside huge areas of nature that we ignore the value, the great importance, of the inspiration and creative experiences afforded by small wildnesses. It is also intended to suggest that, as Thoreau suggested, human beings are part and parcel of wilderness.

Understanding the processes that produce what we consider nature and determining the causes that set those processes in motion are two different things. Thus, if a forest fire of a certain rate and intensity is a process that helps maintain the succession of species over time and thereby allows all the species to persist, then whether the cause of the fire is a person or a lightning strike may be unimportant, and it is certainly a separate issue. We confuse the two. In part, this is because the goal of leaving land "untrammeled" as stated in the Wilderness Act is taken to mean leaving the land without any actual human effect, even though it *could* be taken to mean leaving it without any visible, noticeable human effect.

The state of ecological science is not adequate at this time to resolve the question of how big is big enough to sustain an ecosystem or a set of ecosystems, as proposed in Maine. But the justification given for doing so is unsatisfactory in its present form. I believe that this is because at the base of the human desire to conserve wilderness is the set of intangible benefits people derive from it. It seems that at present, scientific information is insufficient for us to specify an exact size or even the necessary set of habitats for the persistence of many life-forms in wilderness as defined in the Wilderness Act. The human need for an experience of solitude and the intangible benefits that were so important to Thoreau, which many proponents shy away from today, seem a better starting place for justifying the protection of large tracts of wilderness. Ideas have great power, and we need not fear the power of the intangible benefits of the wild as justification for their protection.

Thoreau's writings suggest that he never quite settled in his own mind whether wilderness required vastness, nor did he seem to have faced the question squarely. The preponderance of his writing, however, is consistent with the idea that small wildnesses are enough. It is an issue that modern society seems yet to resolve. My purpose in discussing the proposed Maine Woods National Park and Preserve is to show how Thoreau's approach might be applied to the conservation of nature so that we can deal with what is real, not what we imagine or wish to be true; so that we can save both civilization and nature; so that we can clarify our concepts, separating goals from means; and so that we can succeed in meeting the human needs for both the tangible and intangible qualities of nature through the conservation of specific landscapes.

Viewing the Ocean as Nature

"Like the sea itself, the shore fascinates us who return to it, the place of our dim ancestral beginnings. The edge of the sea is a strange and beautiful place . . . always it remains an elusive and indefinable boundary."

Rachel Carson, *Edge of the Sea*

"It was something formidable and swift, like the sudden smashing of a vial of wrath. It seemed to explode all round the ship with an overpowering concussion and a rush of great waters, as if an immense dam had been blown up to windward. In an instant the men lost touch of each other. There is a disintegrating power of a great wind [in a storm at sea]: it isolates one from one's kind."

Joseph Conrad, "Typhoon"

ALTHOUGH HE MADE four trips to Cape Cod—in 1849, 1850, 1855, and 1857—Henry David Thoreau admitted that he was a "novice" about the ocean. "I have spent, in all, about three weeks on the Cape," he noted at the beginning of his book *Cape Cod*. "I got but little salted," he continued, and "my readers must expect only so much saltness as the land breeze acquires from blowing over an arm of the sea, or is tasted on the windows and the bark of trees twenty miles inland, after September gales."

His lack of familiarity and contact with the ocean—and therefore with the *ocean as nature*—was illustrated when he saw a fox on the Cape. It was more familiar to him than the fish and whales he had encountered there. The fox on the Cape, like the squirrel he saw in Umbazooksus Swamp, personified Thoreau's feelings. "What is the sea to a fox?" he asked. "What could a fox do, looking on the Atlantic?" What could Thoreau, a person of the woods, fields, and town, do, think, or feel when confronted with the ocean? It is a question that raises a deeper issue: can Thoreau's approach be applied to topics about which he knew little? In a way, this is a key test of Thoreau's approach to knowledge about nature. If it can be broadly useful to us, then it should help us even with issues that were unfamiliar to Thoreau, ones that have arisen since his time.

In spite of—or perhaps because of—his unfamiliarity with the Atlantic Ocean, Thoreau found the Cape "a most advantageous point from which to contemplate this world." He found the shore "a sort of neutral ground." It was a "wild, rank place," with "no flattery in it." It treated the dead of all kinds alike, without special sympathy or favor. Its shores were "some vast *morgues,*" he wrote, "strewn with crabs, horse-shoes and razor clams, and whatever the sea casts up," including "the carcasses of men and beasts," which "together lie stately on its shelf, rotting and bleaching in the sun and waves, [as] each tide turns them in their beds, and tucks fresh sand under them."

Thoreau wrote much about the dead on the shore, both unfortunate human travelers drowned in shipwrecks whose bodies had washed onto Cape beaches and animals killed as part of the harvest of natural resources. His book *Cape Cod* opens with a description of the results of the wreck of the *St. John,* a ship carrying passengers from Ireland. The ship had sunk off Cohasset during a violent storm on October 7, 1849, and Thoreau stopped at the beach to view the bodies, bits of clothing, and other personal belongings of those who had drowned. He saw "a man's clothes on a rock" and "a woman's scarf, a gown, a straw bonnet." But "in the very midst of the crowd about this wreck, there were men with carts busily collecting the seaweed which the storm had cast up, and conveying it beyond the reach of the tide," he wrote. "Drown who might," those collecting the seaweed "did not forget that this weed was a valuable manure. This shipwreck had not produced visible vibration in the fabric of society." It was an uncaring and violent nature, this ocean, like Mount Katahdin; life had to continue, and civilization had to produce food and commerce, in spite of nature's effects.

During his third trip to Cape Cod, in July 1855, Thoreau found the carcass of a small whale on a beach near a place he referred to as Great Hollow. Thoreau called the animal a "blackfish," its common name at that time. Today, we would call it a pilot whale. Pilot whales, relatives of porpoises, killer whales, and sperm whales, are small, toothed whales with round, bulging foreheads, slender flippers, and beak-like snouts. They are schooling animals found throughout the world, except in the coldest waters, often in groups of hundreds or thousands.

The blackfish carcass at Great Hollow beach had been there for several weeks. It had been stripped of its blubber and its head had been cut off, but the rest of the flesh remained. Thoreau described it without emotion. He returned "about a week afterward" and saw that the beach "was strewn, as far as I could see with a glass, with the carcasses of blackfish stripped of their blubber and their heads cut off; the latter lying higher up. Walking on the beach was out of the question on account of the stench."

Always one to focus on the complete use of biological resources, Thoreau wrote that the decomposing flesh of the whales "might be made into guano, and Cape Cod is not so fertile that her inhabitants can afford to do without this manure." He had observed the sandy soil of the Cape and knew that it was relatively infertile and, in his typical way, had turned his thoughts to how nature's resources could benefit people. In this case, there were potential benefits, as he saw it, for the harvesters of the whales and for the farming residents of the Cape.

Nearby, he "found a fisherman and some boys on the watch, and counted thirty blackfish, just killed, with many lance wounds, and the water was more or less bloody around." He got a good look at the whales, which he described as "a smooth shining black, like India-rubber," and having "remarkable simple and lumpish forms for animated creatures, with a blunt round snout or head, whale-like, and simple stiff-looking flippers." The largest were about fifteen feet long; the young, without teeth, two feet. "The fisherman slashed one with his jackknife, to show me how thick the blubber was—about three inches," he wrote. "They get commonly a barrel of oil, worth fifteen or twenty dollars, to a fish." Here, we catch a glimpse of Thoreau the practical man of the pencil company, seldom apparent in his writings.

Thoreau reported these experiences without a lament for the hunting or beaching of the whales, nor for the harvest of their blubber, as one would expect today from people with Thoreau's general concern for and sensitivity to nature. In our modern world, marine mammals have become a topic of special concern. The International Whaling Commission meets annually in Cambridge, England, where nations discuss whether to hunt whales

and, if so, how and how many. The United States has the Marine Mammal Protection Act of 1972 as well as organizations, such as Friends of the Sea Otter, devoted to conservation of specific species or groups of marine mammals. Many people are fascinated with the apparent intelligence of marine mammals. Ships take tourists to see whales on both coasts of the United States and along the shores of Hawaii.

Not only did Thoreau report the harvest of whales without lamentation, he described the hunting of these whales—a process of herding them with boats so that they beached themselves and were easily killed—with enthusiasm, as sport and adventure. This seems even more of a contradiction in the context of modern environmentalism. Standing onshore with the whalers, Thoreau heard the "cry of 'another school.'" Looking to the sea, he "could see their black backs and their blowing about a mile northward, as they went leaping over the sea like horses," he wrote, evoking the beauty of these animals as would be done today on television, in the movies, or in a popular article about whales.

He continued:

> Some boats were already in pursuit there driving them toward the beach. Other fishermen and boys running up began to jump into the boat and push them off from where I stood, and I might have gone too had I cho-

> sen. Soon there were twenty-five or thirty boats in pursuit, some large ones under sail, and others rowing with might and main, keeping outside of the school, those nearest to the fishes striking on the sides of their boats and blowing horns to drive them on to the beach.

Then, most interestingly, he noted: "It was an exciting race," in part because "the fishes had turned and were escaping northward toward Provincetown, only occasionally the back of one being seen. So the nearest crews were compelled to strike them, and we saw several boats soon made fast, each to its fish, which, four or five rods ahead, was drawing it like a race-horse straight toward the beach, leaping half out of water, blowing blood and water from its hole, and leaving a streak of foam behind." Thoreau's enthusiasm for the action is unmistakable. "It was just like pictures of whaling which I have seen, and a fisherman told me that it was nearly as dangerous."

Thoreau's reactions were much like those he expressed when he was in the Maine woods and came across the camp of the "explorers"—the two men out in the woods cruising timber, whom he admired and wished he could accompany—an enthusiasm for the adventure involved in harvesting a natural resource. Even the large numbers of whales taken did not distress him, because this harvest represented a livelihood for the people of Cape Cod. "I learned that a few days before this one hundred and eighty blackfish had been driven ashore in one school at Eastham, a little farther south, and that the keeper of Billingsgate Point light went out one morning about the same time and cut his initials on the backs of a large school which had run ashore in the night, and sold his right to them to Provincetown for one thousand dollars, and probably Provincetown made as much more. Another fisherman told me that nineteen years ago three hundred and eighty were driven ashore in one school at Great Hollow," he wrote.

Did his lack of emotional response simply reflect a lack of interest in the ocean and its use? Apparently not, because one thing about the harvest did concern him and led him to expend considerable effort: a lack of information—especially government-provided information—about these whales that could have helped the people of the Cape.

On his return to Concord, Thoreau sought information about the fisheries. He observed that Massachusetts had "risen and thriven by its fisheries." He quoted Timothy Dwight, a famous New Englander who had written *Travels in New England and New York.* Dwight had noted that when the settlement on Cape Cod was new, much before Thoreau's time, a single vessel had arrived at Provincetown with 44,000 codfish and another had come in from the Grand Banks with 56,000 fish, so heavily laden that its main deck was "eight inches under water in calm weather."

Thoreau sought information about the blackfish from the state of Massachusetts. He read reports of the state's zoological surveys and learned that the blackfish had been "rightfully omitted" from the *Report on the Fishes,* "since it is not a fish." Then he read "Emmons' Report of the Mammalia, but was surprised to find that the seals and whales were omitted by him, because he had no opportunity to observe them." Thus, he found nothing in state materials about the pilot whale.

The lack of government information disturbed him:

> Considering how this State has risen and thriven by its fisheries,—the legislature which authorized the Zoological Survey sat under the emblem of a codfish,—that Nantucket and New Bedford are within our limits,—that an early riser may find a thousand or fifteen hundred dollars' worth of blackfish on the shore in a morning,—that the Pilgrims saw the Indians cutting up a black fish on the shore at Eastham, and called a part of that shore 'Grampus Bay,' from the number of blackfish they found there, before they got to Plymouth,—and that from that time to this these fishes have continued to enrich one or two counties almost annually, and that their decaying carcasses were now poisoning the air of one county for more than thirty miles,—I thought it remarkable that neither the popular nor scientific name was to be found in a report on our Mammalia—a catalogue of the productions of our land and water.

Thoreau's statement demonstrates not only a strong interest in the pilot whale as a commercial resource, a livelihood, for the residents of the Cape but also a strong irritation with the failure of government to obtain, keep, and make available information about natural resources of commercial value—a failure to use quantitative information. This failure frustrated Thoreau, just as he had been frustrated, though amused, by the attitude of his neighbors near Walden Pond, who seemed to prefer to talk about the incredible depth of the pond rather than measure it.

Thoreau supported the economic use of the whales. It is clear that he did not lament the loss of the whales or sentimentalize about them, either as individuals or populations. Once again, his focus was on the well-being of people and ways in which nature's resources could be of human benefit. In this way, Thoreau's discussion of the blackfish is consistent with his discussion of timber cruising and harvesting.

For Thoreau, the ocean, like the summit of Mount Katahdin, was nature that did not care for human beings. "There is naked Nature," he wrote of the ocean he saw from the Cape, "inhumanly sincere, wasting no thought on man." The ocean interested Thoreau less than did the land, much as Katahdin was less important to him after his ascent of it than were the swamps of the Maine forests. The ocean was neither the source nor had the imprint of human history and civilization. "The Indians have left no traces on its surface, but it is the same to the civilized man and the savage," he wrote, and as a result, "we do not associate the idea of antiquity with the ocean, nor wonder how it looked a thousand years ago." The ocean also was not, in Thoreau's time, a source of "tonics and barks that brace mankind." Because of this and its lack of connection to human history as he knew it, the ocean did not seem to him to be part of the "wildness" in which was the "preservation of the world." Except for fish to harvest, sell, and eat, it seemed to have little to offer humanity or civilization.

Nor could Thoreau imagine that people could ever have a significant effect on an ocean. "Serpents, bears, hyenas, tigers, rapidly vanish as civilization advances," he wrote, "but the most populous and civilized city cannot scare a shark far from its wharves. It is no further advanced than Singapore, with its tigers, in this respects." This was Thoreau's metaphorical way of saying that although people had the power to greatly alter nature on the land, we did not, nor would in the future, have that power over the ocean.

Perhaps if Thoreau had been born a Polynesian and the ocean were his Concord woods, he would have formed a different attitude. Oceanography was not yet a science in his day, and Thoreau's attitude toward the ocean reflects that ignorance of his time. But more important, his reactions to the ocean show that a *wilderness without the touch of humanity* and its history and effects was timeless and distant and therefore *not of direct interest to him.*

The marshes, woodlands, and shifting dunes of the Cape through which Thoreau walked was a nature closest to his major focus: a nature he could touch and feel. The ocean view from the beach near Great Harbor resonated less with Thoreau. It represented a vast ocean as an unknown and possibly unknowable physical wilderness. It also represented a source of useful products.

Today, four people could look out from that beach and see the ocean with different perspectives. To one, it might be symbolic of the entire Earth, container of the planetary life-supporting and life-containing system called the biosphere. This person might ask, What is the value to us of oceans as nature? Expanding on this question, he might ask, What is the value to us of the biosphere, as nature distant from our daily lives? This is a rationalist but human-centered perspective.

To another person, the killing of pilot whales might evoke a concern with these animals as individuals, animals perceived to be open to suffering. This is a perception of the modern concern with animal welfare. Thoreau expressed essentially nothing about this concern, although today concern with the environment and concern with animal welfare are often viewed as one and the same.

A third person might stand at the strand and lament the ocean's pollution and, by extension, the destructive effects that civilization seems to be wreaking on the entire Earth. This viewpoint symbolizes the anti-rationalist, biosphere-oriented perspective that civilization must give way, that humanity's ability to think and create, and therefore to destroy nature, must be curtailed.

A fourth person might watch the sun setting over the waves and receive a spiritual and creative uplifting; this represents an emotional and spiritual perspective.

These four perspectives on nature are rarely distinguished as separate and different. But they are different. Each can lead to a different set of decisions. Consider, for example, how each of these perspectives would lead to a different ethical and moral approach to the harvest of pilot whales. From the viewpoint of those concerned with animal welfare, the whales would be among those at the top of an ethical pyramid, and the continued existence of each whale would be the goal; harvesting would be stopped. Human needs and desires would be lower in importance, as would a concern with the biosphere.

From the viewpoint of deep ecologists, pilot whales would be much lower in the hierarchy than the Earth's entire life-supporting system. They would be relatively unimportant except to the extent that they are known to affect the workings of the biosphere. But they would rank morally above human beings because whales, although assumed to be sentient, are not capable of developing technology that could destroy the biosphere. People would rank morally at the bottom of the ethical pyramid.

The rationalist of traditional Western civilization would place human beings and their needs at the top, as did Thoreau, and as was done in Western civilization until the last quarter of the twentieth century. The rationalist would, as Thoreau did, discuss the use of seaweed and dead pilot whales as fertilizer; consider whether the ocean could inspire poets, arouse the human spirit; and ask how the ocean affects the persistence of life and the regulation of climate.

The rationalist would also try to learn the extent to which pilot whales play an essential role in the workings of the entire Earth system. He would attempt to determine a sustainable level of harvest, a level that would not decrease the ability of this species to produce an amount that could be harvested at the same level in the future. The rationalist would seek social mechanisms to ensure that the harvest of whales in international waters—part of a global commons—would be limited. Such a person would, in the end, oppose the harvest of pilot whales if no international mechanism could be found to limit the harvest to a level that allowed the species to endure.

As the last consideration suggests, sometimes different perspectives can lead to the same conclusions. British ecologist Sidney Holt, a leading fisheries and marine mammal biologist who took as his life goal the cessation of all whaling, provides a case in point. Holt began his career with a groundbreaking method for mathematical analysis of fisheries harvests, an approach that continues to influence fisheries and marine mammal management. His rational scientific approach, combined with a deep feeling for and interest in whales, led him to conclude that whaling served no useful modern purpose and should be stopped to prevent the extinction of these animals. Even though his starting point was different from those of the deep ecologists (a name not yet invented when Holt began his career) and supporters of animal welfare, and his ethical hierarchy might have differed from theirs, he decided on a goal that would be acceptable to these groups.

For Thoreau, the ocean was truly the "wildest country"; in *Cape Cod,* he wrote that "the ocean is a wilderness reaching round the globe, wilder than a Bengal jungle, and fuller of monsters, washing the very wharves of our cities and the gardens of our sea-side." Thoreau's lack of knowledge about the ocean and the use of its resources led him to grossly underestimate the power of human technology and the will of human beings to alter the ocean and the biosphere.

Even as Thoreau walked on Cape Cod and stared out at the ocean, Yankee whalers were on the high seas, making two- and three-year voyages from nearby New Bedford, Massachusetts, around South America and through the Strait of Magellan; north in the Pacific Ocean to Hawaii for a brief stop for water, food, and other supplies; and then north into the Bering Sea to hunt bowhead whales. The hunt lasted from 1840 to the beginning of World War I. By 1850, the Yankee whalers had taken one-third of the bowheads they would ever catch; by 1860, they had taken two-thirds of what would ever be caught. They were well on their way to diminishing the number of Pacific bowheads from about 20,000 to about 3,000. Most of this harvest and a consequent decline in abundance of bowhead whales took place during the period of Thoreau's four trips to the Cape. But in contrast with his familiarity with science and scientists of his time and his firsthand experience in forests, Thoreau knew little—or at least wrote little—about whaling on the high seas.

He probably would have been surprised to learn that less than 150 years later, whales would be threatened or endangered species; that the ocean would be polluted on a grand scale, with severely modified nearshore waters at essentially every major city in the world; that pollution would come to the very Cape Cod bay that Thoreau viewed when he looked west from Provincetown toward the end of the Cape. We have indeed done what Thoreau said could not happen: we have, symbolically, scared sharks far from our wharves.

Some scientists in Thoreau's time did see beyond the vastness of the oceans, beyond the marshes and woods occupied by foxes and people, to recognize a potential for human-induced, global-scale environmental change. They saw it as if from a distance, through a glass, darkly, but there on the horizon. These contemporaries were capable of a broad-scale perception that Thoreau did not discuss. It was possible, therefore, for a person in his time to have had an awareness of the planetary aspect of nature, of the connections between life and the environment of the whole Earth.

The idea that global-scale climate change could have taken place began to be debated during Thoreau's lifetime. The earliest discussions arose from evidence suggesting former periods of continental-scale glaciations and, therefore, past global climate change that would have significantly changed the distribution of living things. As mentioned earlier, Jean Louis Rodolphe Agassiz was one of the discovers of continental glaciation.

As a glacier melts, rivers form underneath it, carrying debris in strange, meandering channels below the thick layers of ice. At high enough temperatures, the pressure of the ice itself can create such under-ice meltwater rivers. Moving water sorts the particles it carries by size; the faster the water flows, the larger the materials it carries. The under-ice rivers move fast and build a snaking, meandering line of hills, mainly of sandy soil. During periods of slightly warmer weather, when mountain glaciers retreat, these lines of hills are exposed. Called eskers, they are another characteristic of glaciated topography.

In the eighteenth century, glacial moraines, eskers, and similar deposits were observed at low elevations in Europe but were not yet explained. When Agassiz saw debris left by melting mountain glaciers in the Alps, the pieces of a landscape puzzle came together for him—the same effects would have the same causes. Large moraines, eskers, and other glacial formations found on the mainland below the mountains must have been created by gigantic glaciers that moved across whole continents.

But this answer led to a larger puzzle: major climate changes in the past could have caused glaciers to cover vast areas in Europe and Great Britain that were in Agassiz's day, and are today, farms, woodlands, and towns. What could have caused such global climate changes?

Scientists soon recognized that glaciers had once covered vast areas in North America as well. The theory of continental glaciation became generally accepted, although opposition to it appeared in the scientific literature until the very end of the nineteenth century.

A generation before Thoreau, Thomas Jefferson had wondered about the changes that had occurred in distant places, possibly involving the entire Earth. In his 1787 book *Notes on the State of Virginia,* he discussed the recent discovery of seashells in the Andes at 15,000 feet above sea level. He conjectured how this might have come about, discussing various hypotheses in a clear exposition that would make any modern ecological scientist proud. Jefferson considered the hypothesis that the oceans could have covered the Earth's surface to 15,000 feet. He calculated the depth to which water would rise if all the oceans of the world were spread evenly, and he discussed the physics of lifting water. He noted that pumps could raise water only about 35 feet and that the best one could expect is that if ocean water were evenly distributed over the Earth, sea level would rise by about 50 feet. Clearly, this was not a reasonable explanation for seashells at 15,000 feet.

Jefferson considered the possibility, suggested by some, that storms might have tossed the shells to the tops of the Andes. He reviewed information about the height to which storms had been known to toss materials. This, of course, was much lower than 15,000 feet, so he rejected this hypothesis. At the end of his discussion of the matter, Jefferson wrote: "We must be contented to acknowledge, that this great phenomenon is as yet unsolved. Ignorance is preferable to error; and he is less remote from the truth who believes nothing, than he who believes what is wrong."

Jefferson processed information as did Thoreau, according to the scientific method: he carefully and objectively examined the available information, considered specific hypotheses suggested by others in light of the available evidence, and determined where knowledge at the time stood and what one could reasonably conclude.

Thoreau did not involve himself in this topic or the discussion of continental glaciation, even though the latter created a lively debate during his lifetime. His focus remained local and concerned with the connection between human beings and nature, between civilization and nature, here and now.

During his voyage on the *Beagle* between 1831 and 1836, Charles Darwin saw seashells on the summits of mountains in the Andes and then, on returning to the ship, observed an earthquake whose effect was to cause the land to rise relative to the ocean. Darwin thus had information that Jefferson had not had. From this, he inferred that such earthquakes, occurring over a long time, had moved land from the seashore to the summit of the Andes. Later, after his trips to Cape Cod, Thoreau read Darwin's *Voyage of the Beagle,* in which Darwin recounts these experiences. Therefore, in Thoreau's day, not only were shells known to lie at the top of the Andes, but also a reasonable explanation of the phenomenon had been written, and Thoreau had read it.

Even writers in ancient times had speculated about large-scale, even worldwide, environmental effects. Plutarch, for example, wrote in the first century C.E.: "It is by no means for nothing that [the uninhabited parts of the Earth] come to be. . . . The sea gives off gentle exhalations, and the most pleasant winds when summer is at its height are released and dispersed from the uninhabited and frozen region by the snows that are gradually melting there."

In contrast to Thoreau's local and land-based focus, one of the intriguing aspects of Jefferson's and Darwin's analyses is that they applied the scientific method to global questions. These men thought in global terms about life and its connection to the whole Earth. Their imaginations extended beyond the horizon of a fox. Thoreau's focus was in this sense limited, but his method of approach can be used broadly.

In the eighteenth and nineteenth centuries, conjectures about global nature extended from examination of climatic and oceanographic events to active discussions of possible global influences of life—on the Earth's atmosphere and climate; on oceans, soils, and bedrock; on the shape and persistence of the Earth's landforms; on rates of erosion. In the early nineteenth century, geologists began to formulate explanations for the processes of erosion. They began to ask what forces opposed the slow degradation of the land and what created new landforms. A British geologist, Adam Sedgwick, argued that vegetation was the primary force opposing erosion: "By the processes of vegetable life an incalculable mass of solid matter is absorbed, year after year, from the elastic and non-elastic fluids circulating round the Earth." This material, he wrote, is "thrown down upon" the Earth's surface so that "in this *single* operation there is a *vast counterpoise to all* the agents of destruction."

Sir Charles Lyell, one of the fathers of modern geology, dismissed Sedgwick's argument in his 1832 book *Principles of Geology* as "splendid eloquence." Lyell thought that the idea that life might play an important role in large-scale global processes was not worthy of serious consideration, requiring attention only because "such an opinion has been recently advanced by an eminent geologist." Otherwise, he wrote, "we should have deemed it unnecessary to dwell on propositions which appear to us so clear and obvious."

Lyell and his contemporaries had come to recognize that mountains are formed by uplifting generated by forces deep inside the Earth. They did not know the causes of that uplifting force, but they understood that uplifting of continents had had to occur. Mountains and other areas that had been subjected to such uplifting were observed to be generally denuded of soils, implying that they had not retained soils as they rose.

Lyell knew that animals and plants withdrew chemical elements from air, soil, and water, but he argued that their deaths resulted in only a small and very slow return of these chemicals to the environment. If this were true, he wrote, then "if the operation of animal and vegetable life could restore to the general surface of the continents a portion of the elements of those disintegrated rocks, of which such enormous masses are swept down annually into the sea, [then] the effects would have become ere now most striking; and would have constituted one of the most leading features in the structure and composition of our continents." In fact, they did and do, through sedimentary rocks, but Lyell did not understand that.

As a counterexample, Lyell considered the materials washed down annually into the Bay of Bengal, where, as it appeared to him in 1831, "what remains, whether organic or inorganic, will be the measure of the degradation which thousands of torrents in the Himalaya mountains, and many rivers of other parts of India, bring down in a single year." This was an extreme case, with an abundance of vegetation and a high rate of erosion. The forces of physical erosion were so strong, he believed, that the vegetation "can merely be considered as having been in a slight degree *conservative,*" merely retarding erosion, not acting as a constructive agent or an "antagonist power" against erosion. Lyell then generalized from that case to conclude that vegetation could never play an important role in countering erosion.

The primary point here is that discussions about a global environment and a global nature were in progress during Thoreau's lifetime. These, however, did not capture his imagination, which was focused on a nature he could touch, pick up, stare at, contemplate, and feel. He was the fox looking out at the sea, not familiar with anything beyond the sand and vegetation beneath its feet, other land animals, the rolling dunes, the wetlands and streams. As seen in the interests of his contemporaries, a person with Thoreau's sensibilities might have begun to speculate, while viewing the ocean from Cape Cod, about global events, global nature, and the role of human beings in them. But Thoreau did not.

Thoreau missed these great trends. Unfamiliar with oceans, he saw them primarily as a source of products of direct use to people. To him, oceans were unconnected with human history—not, from his perspective, a source of creative inspiration or of contact with the ennobling qualities of human beings that he found within a swamp. To Thoreau, the ocean was as distant, remote, and harsh—and, therefore, of as little value to him as "wilderness"—as was Mount Katahdin. If the oceans were distant as a form of nature, then how much more so was the planet as a whole? Distant enough to be beyond the interest of Thoreau. If this were so, what, then, might be the use of his methods and ideas for us today in dealing not only with the ocean but with the entire Earth as nature?

Chapter 17

Viewing Our Planet as Nature

the hours rise up putting off stars and it is dawn
into the street of the sky light walks scattering poems

E. E. Cummings

"'I wish we could sit down and play with these rocks for a while,' Scott said as they loped back to the Rover. 'Look at those things!' Then, like Bean before him, he stopped to admire the glistening face of a particularly handsome rock. 'They're shiny. Sparkly! Look at all these babies in here. Man!'"

Description of Commander David R. Scott
exploring the moon's surface during the *Apollo 15* mission, 1971

ALONG THE ALLAGASH RIVER, deep in the Maine woods, during the evening of July 28, 1857, Henry David Thoreau sat through an intense thunderstorm, enjoying it. It was near the end of his last trip to the Maine woods. His tent leaked, and he and his companions "huddled together." But Thoreau was rewarded, he noted, by "some of the grandest thunder which I ever heard—rapid peals, round and plump, bang, bang, bang, in succession, like artillery from some fortress in the sky." Joe Polis, his Indian guide, remarked, "It must be good powder."

The storm, Thoreau contemplated, was "all for the benefit of the moose and us." He continued, "I thought it must be a place that the thunder loved, where the lightning practiced to keep its hand in it." This was one of the rare moments when Thoreau lifted his head toward the sky and wrote about phenomena larger than the touch of leaves, the feel of forests, the call of birds, the view of boulders in a fog. Ordinarily, his focus was on organic nature, on objects he could touch and feel. Only when an aspect of nature rocked him "like artillery" did he turn his attention to processes that could affect a large area, that were coupled to a global system. In this way, his view of nature may seem limited to us at the start of the twenty-first century.

Thoreau lived before the creation of one of the most important images of the twentieth century to affect the human spirit and our understanding of the physical properties of nature: the view of the Earth from space as photographed from the Apollo spacecraft in 1969. This image of our small blue planet, mistily obscuring its green-brown continents behind swirling, white water clouds, appears everywhere. In that image, the Earth hangs, as if suspended by some magical force, within the vast, black emptiness of space. Often, it strikes viewers as a blue jewel of space, beautiful enough to decorate the neck of the greatest of goddesses. It is an image of sublime beauty. It has coherence, complexity, and a sense of mystery—of things not yet discovered, of scenery hidden from view. It portrays our loneliness on an Earth seemingly by itself as a life-protecting and life-sustaining entity, alone in the otherwise homeless cosmos. If Thoreau found Mount Katahdin hostile and uncaring toward human beings, imagine how he might have reacted to the view of that black emptiness surrounding planet Earth.

Not having seen the image of the Earth from space or explored global aspects of nature, Thoreau missed one of the major scientific thrusts of our time, the now emerging discipline of Earth system science, which attempts to deal with the planetary—the whole Earth's—life-containing and life-supporting system. This is the new science of global nature.

If Thoreau missed this great trend, of what use can his ideas and approaches be today in dealing with global issues about nature, in determining how to sustain civilization and nature at a global scale and how a person might connect both spiritually and physically to nature at this scale? Perhaps, in a sense, this is an ultimate test of whether Thoreau's approaches can be extended to that which he did not know.

This raises the question, Can an individual relate to something as huge and abstract as global nature in an inspirational, spiritual, or religious way, as Thoreau did on walks to a local swamp or the woods near Concord? We do not yet have a space traveler equivalent to a Thoreau looking down from Mount Katahdin, gazing up from a swamp, or examining the bark of a tree "with microscopic eye." But we do have the words of astronauts, cosmonauts, and scientists who took part in the Apollo missions to the moon, traveled around the Earth in a space shuttle, or viewed the Earth from Russia's Mir Space Station. Their words reveal great reverence, inspiration, a sense of beauty, and emotional reaction to the view of the Earth.

For Russian cosmonaut Aleksei Leonov, the view from space evoked a spiritual feeling: "The Earth was small, light blue, and so touchingly alone, our home that must be defended like a holy relic. The Earth was absolutely round. I believe I never knew what the word *round* meant until I saw Earth from space." Apollo astronaut James Irwin was similarly affected as he traveled to the moon:

> The Earth reminded us of a Christmas tree ornament hanging in the blackness of space. As we got farther and farther away it diminished in size. Finally it shrank to the size of a marble, the most beautiful marble you can imagine. That beautiful, warm, living object looked so fragile, so delicate, that if you touched it with a finger it would crumble and fall apart. Seeing this has to change a man, has to make a man appreciate the creation of God and the love of God.

A sense of beauty and religious inspiration were evoked by a feat of twentieth-century technology. "My view of our planet was a glimpse of divinity," said Edgar Mitchell, who, during the *Apollo 14* mission, rounded the moon and saw the Earth rise "like a small pearl in a thick sea of black mystery."

The image of the Earth from space, now seen so often on book covers, in advertisements, and on television, is a snapshot of an incredibly dynamic system. For U.S. astronaut Charles Walker, that view evoked a sense of musical beauty, of the dynamism of the Earth as a system:

> My first view—a panorama of brilliant deep blue ocean, shot with shades of green and gray and white—was of atolls and clouds. Close to the window I could see that this Pacific scene in motion was rimmed by the great curved limb of the Earth. It had a thin halo of blue held close, and beyond, black space. I held my breath, but something was missing—I felt strangely unfulfilled. Here was a tremendous visual spectacle, but viewed in silence. There was no grand musical accompaniment; no triumphant, inspired sonata or symphony. Each one of us must write the music of this sphere for ourselves.

These experiences tell us that the view from near-Earth orbit evokes a sense of the intangibles—of beauty, creativity, spirituality, and religious feeling. What is striking is that these impressions were not limited to a single astronaut or a single space voyage. If a view of the Earth from space, a view of global nature, can evoke such feelings among reputedly tough, cool, and unemotive pilots and scientists, what might Thoreau find there?

Again and again, those who have seen the Earth from space speak of being awed by its apparent fragility as a life-supporting system and the fragility of the life that fills the Earth's surface. "Before I flew I was already aware of how small and vulnerable our planet is," said Sigmund Jähn of the German Democratic Republic, "but only when I saw it from space, in all its ineffable beauty and fragility, did I realize that human kind's most urgent task is to cherish and preserve it for future generations." Ulf Merbold of the Federal Republic of Germany recalled: "For the first time in my life I saw the horizon as a curved line. It was accentuated by a thin seam of dark blue light—our atmosphere. Obviously this was not the ocean of air I had been told it was so many times in my life. I was terrified by its fragile appearance."

If these cool space travelers received a sense of the fragility of all life, isolated as it appears in its bioaquarium, who cannot wonder that we are living in a time when people fear for the persistence of all life and its planetary life-support system? Experience in space seems to have made some of these explorers into converts for the need to conserve life and its environment. "A Chinese tale tells of some men sent to harm a young girl who, upon seeing her beauty, become her protectors rather than her violators," said U.S. scientist and astronaut Taylor Wang. "That's how I felt seeing the Earth for the first time. I could not help but love and cherish her."

The Earth's fragility is emphasized by the blackness that surrounds it. "Looking outward to the blackness of space, sprinkled with the glory of a universe of lights, I saw majesty—but no welcome. Below was a welcoming planet. There, contained in the thin, moving, incredibly fragile shell of the biosphere is everything that is dear to you, all the human drama and comedy. That's where life is; that's where all the good stuff is," said U.S. scientist and astronaut Loren Acton.

The Earth struck these early near-space explorers as so beautiful, inspiring, and fragile as to raise the question, What is actually required to sustain life on Earth? The question brings up environmental issues that have become major ones since the dawn of the nuclear age and the fear it brought—that a nuclear war might end all human life, perhaps all life. Also evident in the reports sent back by these space travelers is a concern that subtler human-caused global environmental alterations might threaten the planet's ability to sustain life. Some of the statements evoke a sense that we have altered the Earth's environment at a global level. If we have in fact done so, perhaps we have no choice but to understand these changes and take constructive action. We understand little about the requirements for life's persistence, except in the broadest terms.

Another new set of images of the Earth as a global life-supporting and life-containing system provides insights that can help us with this question. These are the images created by twentieth-century discoveries in the fossil record. This wonderful history of the Earth and its life evokes images of past times that no previous civilization, no earlier person, could have known about and appreciated. These images spark our imaginations, creativity, and spirituality. It is a great thrill to imagine ancient landscapes, habitats, and creatures. The large numbers of people who visit dinosaur exhibits in museums, who saw the film *Jurassic Park*, and who read the numerous children's books about dinosaurs and other past life-forms testify to the fossil record's ability to evoke emotions, a sense of beauty, excitement, and adventure, and to inspire human creativity.

This is a truly ancient story that is still unfolding. The history of life on Earth includes a series of major stages, each of which was dominated by certain forms of life and persisted for a very long time in comparison with a human lifetime. We can think of these stages as a set of biosphere "options"—a selection of biospheres each of which "worked" in the sense that it persisted for a long time. Together, they represent a selection of choices rather than a series of "right" or "wrong" states for life on Earth. We can ask, at the global level of concern, what is "natural" and what it might mean to "return" the entire Earth to a natural state. Before modern scientific knowledge about the fossil record, nature seemed to have had one natural state. But the fossil record tells us that there have been a series of stages that existed for so long as to dwarf the time span of what had previously been taken as natural in the usual way the word is used—prior to human effects. If we have altered the environmental globally, and if there are biospheric "options," then we can consider these as alternatives—not so much as ones we can in reality create but as examples whose characteristics may better inform us of what is required for life to persist with the qualities we desire.

The earliest life on Earth occurred about 3.5 billion years ago, at least insofar as the present fossil record tells us. This life consisted of ancient relatives of bacteria, referred to by scientists as prokaryotic life. One striking quality of this first biosphere is the length of time it persisted—2 billion years, longer than any subsequent stage. These early life-forms existed as single cells and as two-dimensional filaments—thin sheets of cells. There were no multicellular organisms with different cell types, with distinct organs, or with complex, three-dimensional structures.

We can imagine that this ancient world had its own kind of beauty. Indeed, the ability to conceive of how it might have looked is a truly twentieth-century and twenty-first century experience. A few images reflecting this earliest biosphere exist today. Visit the hot springs in Yellowstone National Park, for example, and see the bright colors of the bacteria that live in those hot waters—purples, oranges, yellows. Visit a sewage treatment pond and look at the sludge in which bacteria decompose materials. View a lake heavily polluted by phosphorus-containing fertilizers. Take a flight that lands at San Francisco International Airport and look out the window on the landing approach. Look to the west, at the southern end of San Francisco Bay—you will see purple and red salt ponds where bay water is concentrated to produce commercial salt. Bacteria produce these bright colors. Only a few forms of life, including these bacteria, can survive, prevail, and abound in highly saline waters; these are the descendants of the first biosphere.

Such bright colors might have characterized that ancient Earth. Otherwise, our planet between 3.5 billion and 1 billion years ago was as barren as are the moon and Mars today. It might have had its own geological beauty: a starkness like that of an empty desert, perhaps a subject for a Georgia O'Keeffe painting.

Not only did life as an exclusively bacterial world sustain itself longer than any later stage, but also, bacteriologists believe, these life-forms were extremely hardy. It would be difficult, likely impossible, for us to extinguish these forms of life. Even a nuclear war, which would destroy most other life-forms, would not extinguish all bacteria.

Thus, this earliest biosphere lasted an extraordinarily long time and was resilient, not fragile. At first thought, one might consider the continuation of life the most important goal in conserving global nature, and the measure of its success the length of time a biosphere persists. If one were selecting a biosphere that allowed life to persist for the longest possible time, one could do no better than to choose this ancient one. But clearly, this would be an absurd choice. It would be a world without the animals and plants that people associate with wilderness and nature; without conversation, artistic creativity, inspiration; none of the products of civilization: no philosophy or poetry, painting or sculpture; no debates about the value of conserving life; and no creature who is aware of a possibility of its spirituality or is able to propose a moral hierarchy. A movie version of this world, say, *Precambrian Park,* would have little plot, and the action would seem very slow. People would stand beside a hot spring and watch it boil or gather outside a sewage treatment plant and inhale the odors. The hot spring on the edge of town would be low in biological diversity and in human inspiration and spirituality.

The fossil record tells us clearly that long-term persistence of life is not in itself a sufficient goal to consider in deciding what sort of biosphere we want to maintain on the Earth. The question of what biosphere to choose, so to speak, is revealed as a more complex question, one involving the quality of life: what species are present, not merely whether there is any life present. The fascinating fossil record opens our minds to the idea that human desires do matter. The goal in determining optimal biospheric conditions is not simply to sustain life; it is also a matter of human values. A moral imperative to maintain the biosphere is not free of human perceptions, wants, and desires.

Eventually, that first, ancient biosphere changed. During the first 2 billion years of life on Earth, photosynthesis evolved among some of the bacteria. Photosynthesis releases molecular oxygen, which was toxic within the ancient bacterial cells. Ancient photosynthetic bacteria simply dumped this toxic waste outside their cells, in a sense polluting their environment. The first major effect of this global pollution was to oxidize iron dissolved in ocean water. Because oxidized iron is much less soluble in water, this form of iron sank to the bottom of the ocean. A by-product of this process was the formation of major iron ore beds, which we mine today.

Once most of the iron in ocean water had been oxidized, free oxygen produced by photosynthetic bacteria escaped into the atmosphere. This led to a major change in the entire atmosphere, from nonoxidizing to oxidizing; from rich to poor in carbon dioxide; from low to high concentrations of free oxygen. Bacteria that were adapted only to an oyxgenless atmosphere could not survive in this new atmosphere, nor can they do so now.

Today, bacteria that cannot survive in high-oxygen environments, called anaerobic bacteria, persist in special habitats: in the intestines of ruminants—cows, impalas, elk, moose, and their relatives—where they enable the digestion of grasses, herbs, twigs, and leaves of woody plants; in the innards of termites, where they enable the digestion of wood; and in parts of wetlands where the soil is saturated by water or there is standing water that contains little oxygen, or in other soils where water and particles keep oxygen concentrations low. There, these bacteria continue to perform chemical reactions essential to all life, such as converting free nitrogen in the atmosphere to compounds that all other kinds of life can use and converting nitrogen oxides to free, molecular nitrogen that then returns to the atmosphere. For any biosphere that we know of to persist, there must be oxygenless pockets where such bacteria can survive and where anaerobic bacterial-induced chemical reactions can take place. The rest of us depend on them. Although we would not select a world made up only of these bacteria, we and the rest of life cannot survive without them.

Meanwhile, in that ancient time, some bacterial species evolved that could live in a high-oxygen atmosphere. About a billion years ago, the new, oxygen-rich atmosphere presented new habitats and new opportunities for life, even though it was toxic to the old forms. With easily available free oxygen, organisms could grow larger, be more complex, and use energy faster. They could move quickly.

About 700 million years ago, animals, plants, and fungi began to evolve. The cells making up these forms of life differ from those of bacteria and are much more complex in structure, with their deoxyribonucleic acid (DNA), the genetic material, concentrated in a nucleus rather than distributed throughout the cell, and with organelles such as mitochondria, which process energy.

Diversity of life in the oceans increased rapidly about 570 million years ago. In a cliff of ancient rocks formed during that time, this can be clearly seen as a sharp boundary, with fossils of complex organisms above but no complex organisms below. Known as the beginning of the Cambrian period, this time of transition was characterized by a second type of biosphere that persisted for a long time—from about 570 million to 500 million years ago. During this period, animal life was restricted to oceans—probably quiet bays rather than the open ocean. Biological diversity was low compared with that of many subsequent periods, including our own. Land above the sea remained lifeless, a barren landscape silent of song, empty of green plants and bright flowers and sentient creatures. We can imagine this world with abundant life in quiet bays, perhaps with fewer of the bright colors that characterized the ancient bacterial world. Who would opt for this biosphere as a goal? It would be a world without the nature that so appeals to people.

About 450 million years ago, life began to move onto the land. The first life-forms on land were plants, and the first land animals were early relatives of insects. There was a long period during which no animals with backbones lived on the land. Plants of this time reproduced by spores. The male reproductive cells required water in which to swim to female cells and to produce fertile spores. As a result, these plants were restricted to stream and wetland habitats, where open water was present enough of the time for reproduction to take place. Dry uplands where today we find vast forests, grasslands, and deserts were empty of life, emptier than are the driest parts of Arizona, northern Africa, or central Australia today. The land scape would have been lonely. Over time, tree-like plants developed in wetlands—the ancestors of ground pines, small plants with conifer-like needles that spread along shaded, moist forest floors. Thoreau saw these in the Maine woods and the woods around Concord.

The first ancestors of conifers, ancient gymnosperms, evolved about 400 million years ago, during the middle Devonian period. During that time, all the continents were connected in one supercontinent, referred to today as Pangaea. Gymnosperms were the first plants to evolve pollen, a protective structure containing the male reproductive cell in a waterproof case. For the first time, plants were freed from having to live near streams or wetlands. Gymnosperms expanded into many of the upland areas. The landscape of this time would appear more familiar to us today, more like the ones human beings have come to know, but it was without flowers, grasses, and the vast herds of large mammals that depend on grasses for their food.

Animals with backbones eventually made their way onto the land, but there was a period of approximately 100 million years between the first appearance of such animals and the evolution of vertebrates that could feed on plants. The earliest land animals with backbones either fed on plant-eating insects, were carnivores of vertebrates that fed on insects, were carrion feeders, or fed on marine life but moved about on the land, much as alligators and crocodiles do today. Land vertebrates were amphibians; they required open water to reproduce.

About 360 million years ago, during the Carboniferous period, reptiles evolved watertight skin and eggs with protective, waterproof shells that could survive on land. With these, they could live in upland habitats.

A third option for a sustaining biosphere would be this one, with conifers and reptiles able to live away from open water. But it would still be without flowers and sentient creatures.

Some 200 million years ago, during the Jurassic period, dinosaurs, mammals, and birds evolved and occupied much of the land. Species diversity increased greatly. With the current fascination with dinosaurs, perhaps some would see this type of biosphere as an acceptable goal. However, if the evidence is correct that an asteroid colliding with the Earth dealt the final blow to the dinosaurs and much of the biological diversity of that time, this biosphere was vulnerable to very rare but extreme catastrophes.

Modern space technology appears to offer a way to deflect asteroids that are on a collision path with the Earth and therefore to enable life to protect itself from this rare destructive event. If this can indeed be achieved, it will be a product of human civilization—the result of a long history of creativity, of invention, of wondering about nature, of engineering design, of human imagination. It will have required skills like those of Thoreau the pencil engineer and Thoreau the naturalist. Even if one were to accept the age of dinosaurs as an acceptable goal, its biosphere would be more vulnerable than that possible for us today—and it would still be a biosphere without art and poetry, without organisms capable of a sense of spirituality. It would be a world without civilization.

About 60 million years ago, grasses evolved during what geologists call the Tertiary period. Grasses, able to live on dry steppes, created savannas and prairies, bringing a beautiful bounty of vegetation to large areas of the Earth that were previously sparsely occupied. Along with the grasses and probably in a symbiotic relationship with them, large grazing mammals evolved. Subjected to this grazing, grasses evolved to withstand it. The grasses that survived best under grazing pressure persisted and dominated, and in turn the large grazing mammals that could survive with these new forms of vegetation persisted. This was the setting of the biosphere at the time our species evolved: a world of flowers, grasses, trees adapted to all climates, and a wealth of mammals that grazed and browsed or fed on those that did. This was the pre-Pleistocene biosphere, about 2 million years ago.

I believe that this is one of the three images in the minds of those who argue for a sustained biosphere as the primary moral directive or just argue that we must return nature back to what is "natural." This biosphere would have much of the scenery of the modern world, but it would lack human beings and therefore would not be a place of conscious inspiration, creativity, or spirituality, nor of the rationality that some fear so much as a potentially destructive force against nature.

The second such image, I believe, is that of the world just prior to the development of farming. Yet by this time, human beings, as hunters and gatherers, had greatly affected habitats and species diversity through their use of fire, through the spread of species into new habitats as early peoples migrated around the Earth, and through their hunting of animals. It was a world of growing alteration by human beings. It was also the first biosphere with art and technology.

To make this stage the goal would mean that human inventions, art, and technology, once started, could be held in check, with no further development. Would that be possible, and if so, would it be desirable? From our modern perspective, human innovation, art, and technology might have changed slowly at first, but they changed nonetheless. A world with people who have hunting tools, art, and spirituality is not likely to remain a steady-state world over many millions of years. Inventiveness and creativity, once present, are not easily suppressed. Human culture and civilization remain static only under oppressive political and social systems, and even then only temporarily from a geologic time perspective. The modern idea of individual freedom would seem antithetical to maintaining this as a static biosphere.

Anthropologist Paul S. Martin and his colleagues propose that even at this stage in human development, our species may have caused some major extinctions through the hunting of large mammals. They argue that the baseline conditions for land planning should be those of about 10,000 years ago, at the end of the last major continental glaciation and prior to the extinction of many of what nineteenth-century biologist Alfred Russel Wallace called the biggest, hugest, and fiercest mammals that ever lived. This is the biosphere to which we should seek to return, Martin and his colleagues argue. As a start in doing so, they suggest we bring African elephants to North America to replace the mammoths and mastodons that became extinct soon after the end of the ice ages. But this would be a world without writing. Our species would be present to enjoy nature, to be inspired, to have some creativity stimulated, and to begin to alter the environment on a very large scale. In the suggestion that we bring back the elephants to North America, there is a poignant longing for a golden age long past.

The third type of biosphere that I believe is in the minds of those who argue for a sustained biosphere as the primary moral directive or just argue that we need to return nature to something "natural" is that of the Earth just prior to European settlement of the Americas, Australia, and New Zealand: a medieval or early Renaissance biosphere. The appeal of this short period is that vast areas had not yet been influenced by Western civilization, the source of so much of the technology perceived by some as only destructive to nature. This would leave some who long for a past golden age of nature uncomfortable because farming, large-scale constructions, and development of cities had already begun to affect the Earth. Geologically, this stage did not persist for a long time. It would be highly unlikely for it to persist without change because of the momentum of innovation and creativity. For some, the choice might be an even earlier period—the time before writing, when the only history was oral, when poetry was song. Patents could not be written; technological change was slow. The earliest known alphabet, shown as writings on soft stone cliffs on an ancient road west of the Nile in Egypt, was carved between 1900 and 1800 B.C.E., so this would place the desired time as prior to 1900 B.C.E.

One hears frequently that biological diversity must be maintained as much as possible to allow for the persistence of the biosphere, yet the longest-lasting biosphere, the first, was the least diverse, consisting only of bacteria. At the global level, some forms of life have persisted for very long periods. For those who want to maintain as much biological diversity as possible, it is therefore dangerous to equate maximum species diversity with the maximum likelihood of persistence of the biosphere. If persistence of the system is the goal, the optimal level of diversity is much more likely to be less than the maximum. Once again, the fossil record shows us that what may seem to be simple questions are not so simple.

That there might be an optimal level of biological diversity for the persistence of the biosphere is not a justification for the high rate of species extinction that modern civilization is causing. The possibility that maximum diversity may not be linked to long-term persistence does not mean that a minimum diversity is desirable. Redundancy of function is valuable for the persistence of a complex system, and a high level of diversity can provide such redundancy. We can choose as our goal the maximization of diversity and minimization of extinctions on moral, aesthetic, religious, or inspirational grounds, even if this meant accepting a less stable biosphere. But this is a human choice.

The wonderful images we can construct today from the fossil record suggest that the earlier, rather mechanistic analysis of global environmental issues is an insufficient basis for an ethics regarding nature on a global scale and human relationships with nature. It does not follow that this history leaves us free to destroy life wantonly, without considerable study or care, or provides an excuse to allow the extinction of species. The history of physical processes—the history of the biosphere—tells us what has been possible. The intangible values—spiritual values, a sense of beauty, creative inspiration, religious values—provide a basis for us to select what we desire from the set of what is physically possible. Perhaps the main point is that two very different modern technologies—space travel and paleontology—have opened new windows on nature that stimulate our imagination, creativity, and spirituality and inform us about what is possible. Science and technology have, at a global level, begun to influence our understanding of what choices we have in selecting values.

This global perspective brings us back to the human relationship with nature. From the perspective of the functioning of the biosphere, human beings appear to have long been making alterations *in human terms,* however brief their effect in geologic time. From the discovery of fire and the transport of exotic species from one location to another, early human beings altered the biosphere. How, within this context, do we find an appropriate place and role for humanity and civilization? Using Thoreau's approach, we would begin with the details of new knowledge and take assertions by experts as starting points for further examination; listen to both experiential experts (in this case, the astronauts looking at the Earth) and professional experts (paleontologists, ecologists, and other Earth scientists studying the biosphere as a system); and integrate intangible values with knowledge of physical phenomena. Using this approach, we would struggle to get our values, desires, and choices clearly in mind.

One implication of the foregoing discussion is that it is easier to justify the conservation of biological diversity on aesthetic, creative, inspirational, or moral grounds, making use of scientific knowledge as background, than solely on the basis of the history of life and the mechanistic requirements of the biosphere. This discussion also emphasizes the value of a scientific understanding of how the biosphere functions as background to these values. The discussion suggests that the intangible aspects of the human relationship with nature can provide a valid rationale for conserving species and overall biological diversity. In this way, Thoreau's approach to knowledge, in which human desires regarding both the tangible and the intangible are accepted and understood, can help us arrive at justifications for the conservation of species.

The view of the Earth from space has inspired the imagination and the spiritual and religious feelings in astronauts. Their experiences as local, or experiential, experts—in the sense that they are not professional ecologists but people viewing the Earth from a perspective not available to the rest of us—suggest that seeing the Earth from space brings about an experience of beauty, creativity, and inspiration. A new sense of reverence for life arises, along with new rationales for its conservation. David Scott's exclamations on seeing the rocks on the moon—"Look at those things! . . . They're shiny. Sparkly! Look at all these babies in here. Man!"—suggest that a visit there might even be fun. Deriving a moral perspective from reverence, love of creativity, inspiration, and joy allows for flexibility.

Scientific information is used well in the way that Thoreau used it: to learn how nature works so we know what is possible and we understand how human beings fit into the mechanistic analysis of the biosphere. At the same time one increases this scientific understanding, one can also increase the appreciation of nature's intangible qualities. Science and technology were required to put astronauts into space and on the moon, where they then experienced a sense of beauty, fun, and an obligation to conserve the living Earth.

With the limited scientific knowledge of his time, Thoreau did not have a global perspective on life, nor did he recognize the potential of technology to affect the entire biosphere. However, we can apply his methods to evaluate various justifications for conserving life at a global level. When one reflects on these, it seems clear that the question is not one of a single truth (what is *the* state of the biosphere that can and will persist) but rather a question of *options.* The biosphere has existed in many stages and states, each persisting for tens of millions of years, hundreds of millions of years, or several billion years. We could choose any as our goal as long as we are informed.

It is worthwhile to sum up the desires of some who have become so frustrated with the negative effects of modern civilization that they wish to return to a past golden age. Again, I believe that this desire focuses on one of three past times. The first is more recent than 60 million years ago but before 2 million years ago, when our species evolved—the golden age of grasslands, large grazing and browsing mammals, and these animals' large mammalian predators, unaffected by human actions. It is like an idealized television program about the Serengeti Plain: bountiful, flowering grasslands, forests, and wetlands filled, at least in our imaginations, with millions of beautiful animals quietly going about their business within a great chain of being.

The second golden age is a world with people incapable of altering the Earth—at least before farming and perhaps before the invention of writing, which would place it somewhere between 5,000–10,000 years ago and 2,000 years ago. Here, the wish is for a landscape with people who appreciate the Earth's beauty but have no capacity to destroy it, perhaps no capacity to affect it to any significant degree. Spirituality, creativity, and inspiration would have their place, as we know from cave paintings and other artifacts. The implicit assumption is that such human activities would remain in check, that creativity would never lead to new technologies. But is this realistic? Once people became capable of imagination, of art, even of the most primitive toolmaking, perhaps a process was set in motion from which there was no turning back. People simply will not stop being creative, even though they may be forced to do so for a time. And even if most people are uncreative most of the time, there will always be somebody with a new idea who cannot wait to try it out. If this is true, then once the Earth was peopled, there was no longer a possibility of the kind of steady-state "balance of nature" that is such an important myth in Western history. As pointed out earlier, a society in which technological innovation is held in check would be a society of rigid rules in which individuality is suppressed—a society most unlike a modern democracy.

The third golden age in many people's minds is, I believe, the world before European expansion, when large areas were not yet affected by Western technology. Some might say that if only civilization could have remained at that stage of development, large areas of what we today consider wilderness would persist. But as with the second golden age, this is a highly unlikely scenario. People are just too inventive, and there are always independent oddballs who will try to change things. So it would seem that none of these golden ages is actually attainable—and even if any were, few people today would choose them because doing so would require that we cease to innovate.

The history of the biosphere as revealed by twentieth-century scientific study of the fossil record suggests that we really are stuck with people and nature as one dynamic, changing system, even if we do not like it. If Thoreau were here, I believe that he would like it.

The marvelous scientific advances of our age—new perspectives of the Earth gained from space travel, of the biosphere as a global system, of the history of life and its environment on the Earth—can open up many positive and constructive ideas about people and nature if only we allow them to. This is not to deny the importance of global environmental problems and our need to deal with them. It is to suggest that we must face these realistically, accepting the fact that human innovation will continue and that we are unlikely to return willingly to a very small human population.

Much of the new scientific work to understand the biosphere as a functioning system began in the last three decades of the twentieth century. This work has included the search to develop global models of climate, of the oceans, and of the Earth's vegetation and to understand how life is interconnected at the global level. From the work of G. Evelyn Hutchinson to that of James Lovelock, from the earliest attempt to use formal mathematics in forecasting the weather, around the time of World War I, to the computer simulations of today, scientists are increasing our understanding of how this global system functions. It is just a start, and the problems humanity has created are outrunning our understanding of what to do about them. Another realization that comes from a global perspective is that we have a responsibility for humanity and nature at the global level, and that to wish for the return of a golden age, however understandable that may be, is not to face up to that responsibility.

We need to consider detailed information about the Earth's life-supporting system, following the process Thoreau did when he studied the spread of seeds, kept track of seasonal changes in the plants he revisited on his daily walks, and measured the depth of Walden Pond. The decisions we face are not free of human values. Civilization and nature even at a global level are linked together, just as Thoreau saw them locally, around Concord, and to save them we must approach them as part of the same process.

Conserving Mono Lake: Walden Pond as Metaphor

"Now the elements of the art of war are first, measurement of space; second, estimation of quantities; third, calculations; fourth, comparisons; and fifth, chances of victory. Measurements of space are derived from the ground. . . . Quantities derive from measurement, figures from quantities, comparisons from figures, and victory from comparisons."

Sun-tzu, *The Art of War*

"The islands in [Mono Lake] being merely huge masses of lava, coated over with ashes and pumice-stone, and utterly innocent of vegetation or anything that would burn; and sea-gulls' eggs being entirely useless to anybody unless they be cooked, Nature has provided an unfailing spring of boiling water on the largest island, and you can put your eggs in there, and in four minutes you can boil them as hard as any statement I have made during the past fifteen years. Within ten feet of the boiling spring is a spring of pure cold water, sweet and wholesome. So, in that island you get your board and washing free of charge—and if nature had gone further and furnished a nice American hotel clerk who was crusty and disobliging, and didn't know anything about the time tables, or the railroad routes, or about anything—and was proud of it—I would not wish for a more desirable boarding-house."

Mark Twain, *Roughing It*

THE STEPS Henry David Thoreau followed in determining the depth of Walden Pond and using that information to formulate generalizations applicable to other lakes are the steps of the scientific method. Although the scientific method is assumed to be the basis for solving current environmental problems, it is, in my experience, used less commonly than is generally supposed. Does the fact that Thoreau used the scientific method make any difference to the crowds of people who visit Walden Pond each year? What do the depth and shape of the pond matter to them? How would any generalizations arrived at by Thoreau affect their lives or their enjoyment of Walden Pond? Why should they care about the scientific study of lakes?

Perhaps these questions are best answered by an example. One hundred forty years after Thoreau ruminated on lakes in Concord, Massachusetts, the depth and shape of the bottom of Mono Lake in California were the key to solving a modern environmental controversy there. Mono Lake is just east of the Sierra Nevada and the major peaks of Yosemite valley and is near the route traveled by Edwin Bryant.

Mono Lake is salty because of the great climate changes that have taken place since the last ice age. At the end of this ice age, when the climate in eastern California was wetter, Mono Lake was formed when freshwater flowed off the Sierra Nevada and into a large basin. This water continued to flow for thousands of years, evaporating in the hot desert sun and leaving behind its dissolved chemical elements. By the twentieth century, the lake was more than twice as salty as the ocean and highly alkaline, inhabited by algae and bacteria that could withstand these harsh conditions. A brine shrimp and the larvae of a brine fly, two small animals adapted to salt lakes, fed on the algae and bacteria. People who have home aquariums are familiar with brine shrimp as a food for pet fish; close relatives of these shrimp from salt ponds in San Francisco Bay are an important food for aquaculture around the world. The brine shrimp, brine fly larvae, and other organisms able to survive in Mono Lake grew in great abundance and were highly productive.

The shrimp and fly larvae provided abundant food for five bird species: the California gull, the eared grebe, the snowy plover, and two phalaropes—Wilson's and red-necked. About 1.3 million birds nested along the lake or stopped there to feed during migration. The food the migrating birds obtained was essential fuel for them to continue their flights.

In the 1940s, the city of Los Angeles began to divert all stream water that had previously flowed into Mono Lake to augment the city's water supply. Divested of its major source of water, the lake began to dry up. It covered 60,000 acres in the 1940s, but by the 1980s, the lake had lost one-third of its area and covered only 40,000 acres. In the mid-1970s, students at the University of California, Berkeley, became sufficiently concerned about Mono Lake to organize the Mono Lake Committee. The National Audubon Society helped raise funds, and a San Francisco law firm took on the case pro bono. The Mono Lake Committee established offices in Los Angeles and Lee Vining, a small town on the edge of the lake. A dynamic and able director, Martha Davis, pursued conservation of the lake in the courts, lobbied the state legislature, negotiated with the city of Los Angeles, and publicized the issue. "Save Mono Lake" was commonly seen on bumper stickers in the Golden State.

The debate over the lake focused on one issue: without the supply of surface water that had formerly flowed into it, would the lake dry out to the point of becoming so salty and alkaline that all life would stop and all the lake's aesthetic values would be lost? Alternatively, would the lake reach a steady state, as the issue was expressed at the time, balancing surface evaporation with rainfall and subsurface flows sufficiently to sustain this unusual ecosystem?

The Mono Lake Committee argued that the steady-state condition would be one in which algae, bacteria, brine flies, and brine shrimp would die and birds would cease to have a supply of food; that the water level at this steady state would be so low that the islands in the lake, used as nesting areas by California gulls, would become peninsulas, exposing the birds to attack by coyotes. The Los Angeles Department of Water and Power responded that the steady-state water level would be higher than that and that rainfall and subsurface flows would support the lake's life and its populations of birds. Resolution of the conflict required knowledge of the depth of the lake and the topography of the lake basin, the same questions that had so intrigued Thoreau about Walden Pond and other lakes near Concord, Massachusetts.

In 1984, as a result of lobbying by the Mono Lake Committee, the state of California passed A.B. 1614, an assembly bill providing $250,000 for scientific studies of "the effects of water diversions on the Mono Lake ecosystem" and stipulating that "the study shall evaluate the effects of declining lake levels, increasing salinity, and other limnological changes of Mono Lake" on "the total productivity, seasonality and physiology of brine shrimp, flies, and algae living in and around Mono Lake"; on "the numbers, productivity, physiology and residency patterns of breeding and migratory bird populations"; on dust storms resulting from the widening sand shores as the lake area decreased; and on the hydrology of the lake, its evaporation, and freshwater spring flow.

Originally, the California Department of Fish and Game was to conduct the study, but in an apparent move to avoid the controversial topic, the department approached the University of California, which in turn asked me to direct the study. I convened a small panel of experts from around the country, each representing a discipline important to the issues: Wallace Broecker, a geochemist; Lorne Everett, a hydrologist; Joseph Shapiro, a limnologist; and John Wiens, an ecologist specializing in the study of birds.

Our approach was to begin by examining available scientific information about the lake. Prior research at Mono Lake had primarily involved scientific observation and description, not a search for generalizations. Two scientists had studied the birds of the lake for years, but much of their work remained unpublished. The panel decided to allocate much of the funding to the people who had been studying the lake so that they could summarize and generalize their results. The panel then considered the implications of this information. Work performed under the direction of John Melack of the University of California, Santa Barbara, provided information about the salinity levels at which the lake's algae, brine flies, and brine shrimp would no longer be able to complete their life cycles. If this were to happen, the lake would "die" as an integrated ecosystem; it would cease to sustain life. In particular, it would cease to sustain the life that was food for the more than one million birds that visited or nested at the lake.

This led to a question: When, if ever, would the lake reach this salinity? As mentioned earlier, the Los Angeles Department of Water and Power argued that rainfall and subsurface flows would balance the evaporation at a water level sufficient to sustain life in the lake. However, the Mono Lake Committee argued that the lake would dry out more than that, to a point at which all life would cease.

Resolution of the issue required two kinds of scientific understanding. One was knowledge of evaporation as a function of the lake's area and the local climate, but a method for calculating this did not exist. The second was knowledge of the lake's water volume so that salinity changes could be calculated for each lake level. This required a map of the lake showing its depth and the shape of its basin, exactly as Thoreau had sought for Walden

Pond. As with Walden Pond in Thoreau's time, this information did not exist. No one had mapped the basin of Mono Lake.

We allocated funds to develop a model for calculation of evaporation and, taking a step into the twentieth century, hired a firm that used sonar to make a complete map of the lake basin. It seemed that little had changed in human behavior toward ponds and lakes in a century and a half. Just as people in Thoreau's day had speculated about the depth of Walden Pond but had not bothered to measure it, people in our time had speculated about the fate of Mono Lake but had not made the scientific measurements required to resolve the controversy. This statewide issue of considerable popular concern had been debated for more than a decade but could not be resolved without these two pieces of information. Why had they not been obtained? In part, the need to know this information had become clear only after we reviewed all the available information and thought through the implications. The process of careful analysis and reflection on observations had revealed what additional information was needed.

From the scientific analyses, it became clear that there were three crucial lake levels, and these we presented to the public as options from which to choose, depending on society's desires for the lake. At the highest lake level, all environmental values would be protected: birds' nesting habitats and food; the scenic beauty of this large lake beneath high mountains; and the strange, sculptural rock formations called tufa towers that were produced in the lake but were eroding away as the water evaporated. Tufa is a soft rock-like material formed when volcanic gases bubble through highly saline and alkaline waters. As the lake had gradually become drier and shallower after the last ice age, many of the tufa towers had been exposed. The soft minerals that form tufa are easily eroded by waves, and the declining lake had undermined the towers, which had toppled in greater numbers as the years passed. Keeping the water level high would preserve most of the remaining towers.

At the intermediate lake level, some bird nesting habitat would be reduced and the water level would drop low enough for coyotes to cross over to an island where gulls nested, reducing gull reproduction. At this level, the lake would lose some of its aesthetic qualities but would continue to function as an ecosystem.

At the lowest level, the lake would continue as a functioning ecosystem but would retain little else for which it was appreciated. Below this level, the lake would become too saline and alkaline for the brine shrimp, brine flies, and algae to survive, and the lake would die.

These options were presented to the public, and the Mono Lake Committee began negotiations with the Los Angeles Department of Water and Power. Before our study, a court had determined that the city of Los Angeles could continue to divert all the water that had previously flowed into the lake unless it could be demonstrated that this diversion had negative environmental effects. After the study, another court determined that the lake level had to be returned to the highest of the three levels and that all water diversions must stop until either the highest level was reached or the city could demonstrate that some lower level would not have a negative environmental effect. After several years of negotiations, the city of Los Angeles formally gave up all rights to the water that flowed into Mono Lake. Thus, the pursuit of knowledge of the lake's bottom topography and other scientific information changed the future of the lake.

Before the study began, I had never visited Mono Lake. One reason for my involvement in the project was that this lack of previous contact allowed me to be independent of biases. During the course of the project, my colleagues and I made a number of visits to the lake. In scientific work, one abstracts one's reactions to the environment and the task at hand; whether one enjoys or hates it is irrelevant. It was always necessary to detach myself when working on this study from any sense of liking or disliking the lake in order to analyze the information in a neutral, independent, professional manner.

Be that as it may, I found Mono Lake a kind of magical place. Set in the high desert, it is surrounded by typical open desert shrubland; sagebrush, coyote bush, and other aromatic plants scent the dry air. In that dry climate, the horizon telescopes. To the west stand the magnificent, snowy peaks of the Sierra Nevada, with streaks of white, dark green, and light green extending down the slopes like runny watercolors. As we hiked on the eastern side of the lake, past the strange towers of tufa stone, midday heat made the air undulate. There is a certain wild quality to this open desert country, which is sparse in vegetation and often disturbed by drought and downpour. It is a silent country of spicy aromas. The time I spent at Mono Lake affected my inner sense of nature, teaching me about another natural area. For me, doing science did not destroy the spiritual experience of nature; likewise, the inner experience of nature did not interfere with my scientific analysis. I had found a use of science—helping to solve the controversy over Mono Lake—and at the same time I had deepened my personal connection with nature. In retrospect, I realize that my colleagues and I applied the same methods Thoreau had: seeking to find and understand what quantitative information was already available and to apply the scientific method; seeking the ideas of experts but subjecting them to additional tests and analyses; seeking to maintain our appreciation of nature as beautiful and inspirational but not to impose ideologies, or "expectations," in Thoreau's terms, on our attempt to understand the functioning of Mono Lake. Taking this approach, we had solved an environmental problem.

To return to the questions posed at the beginning of this chapter, our use of the scientific method—similar to Thoreau's use of it at Walden Pond—made a great difference to the crowds of people who value Mono Lake. Even if they are not aware of it, the lake's depth and shape matter to them because these held the key to understanding the lake's functioning as an ecosystem. Our search for generalizations, once again following Thoreau's method, affected the lives of California residents both in terms of their ability to appreciate the beauty of Mono Lake and in terms of changes in water supply and quality. Thus, the Mono Lake study demonstrates how Thoreau's scientific approach can help us balance the competing interests of nature and civilization: the city of Los Angeles continues to function, and so does Mono Lake.

Cities, Civilization, and Nature

Earth has not anything to show more fair:
Dull would he be of soul who could pass by
A sight so touching in its majesty:
This City now doth like a garment wear
The beauty of the morning; silent, bare,
Ships, towers, Domes, theatres and temples lie
Open unto the fields and to the sky;
All bright and beautiful in the smokeless air,
Never did sun more beautifully steep
In his first Splendour valley, rock or hill.
William Wordsworth, "Upon Westminster Bridge"

"We require an infusion of hemlock spruce or arbor-vitae in our tea."
Henry David Thoreau, "Walking"

ALTHOUGH THE IMAGE of Henry David Thoreau living alone in a cabin he had built by himself at Walden Pond suggests that he was a loner, and the fact that he never married and spent much time alone suggests that he was almost a hermit, disliking the society of other people, this was not the case. Thoreau never abandoned the city nor what we today call the suburbs. During the time that he lived at Walden Pond, he went into town almost every day, had friends in cities, and entertained many guests in his cabin on the lake. "I think that I love society as much as most," Thoreau commented at the beginning of his chapter in *Walden* titled "Visitors." "I am naturally no hermit, but might possibly sit out the sturdiest frequenter of the bar-room, if my business called me thither." Thoreau disdained trivial society and conversation, but he sought meaningful conversations at Walden Pond, just as when he sought to understand Joe Polis in the Maine woods and the Wellfleet Oysterman and the keeper of the Highland Light on Cape Cod.

"I had more visitors while I lived in the wood than at any other period of my life; I mean that I had some," Thoreau wrote, with his typically wry humor. "I met several there under more favorable circumstances than I could anywhere else. But fewer came to see me upon trivial business. In this respect, my company was winnowed by my mere distance from town. I had withdrawn so far within the great ocean of solitude, into which the rivers of society empty, that for the most part, so far as my needs were concerned, only the finest sediment was deposited around me."

For most of his life he was a resident of Concord, and in that sense he was a suburban or urban person with an intense interest in nature. Thus, Thoreau's life brings up the question, What is the role of cities in an environment that is both spiritually and physically meaningful to people? In present-day terms, the question might be phrased, Is there any way that a city could have a place in a "good" environment? This takes us to the heart of the question: Are nature and modern civilization compatible?

These questions are especially poignant because most of us live in cities and admire the countryside. There is much prejudice against cities today. If we think about wilderness at all, we admire it, but we rarely visit it. Our idea of nature is molded by television programs and color photographs in magazines. When we focus on the idea of nature, it is a focus on the environment of our vacations and of places we rarely experience directly. This leads to a peculiar idea about nature and our personal relationship with it. It is a vicarious relationship, maintained secondhand through the pronouncements of sonorous voice-overs as we watch lions hunting impalas and hawks hunting field mice and rabbits on a color screen in the family room. This results in a false dualism: nature is out there, to be admired, preserved, revered; life is here, in the home or on the curb or where the tarmac runs into the mud. We tend to miss the point that the environment that influences us most is the one with which we are in contact most of the time.

For most of us, nature is an urban or suburban environment. If we are to come to terms with nature and have direct contact with it with any frequency, to become familiar with it as Thoreau did, then we must seek that nature in our cities and suburbs. If we believe that a harmony between people and nature is necessary for ourselves, for civilization, and for life on Earth, then we must seek this harmony in cities as well as in the countryside. If there are certain essential qualities to a nature suited to human beings, we must seek to establish those in cities. We must integrate nature into the city and the city into nature; we need Thoreau's swamp on the edge of town. The path to accomplishing this is not so much to look "out there" beyond the cityscape as it is to design our cities so that nature exists next door and on the doorstep. Such designs can not only serve the spiritual needs of people but also conserve biological diversity in ways little appreciated except by experts on conservation ecology and landscape architecture.

We must focus on cities as nature. It has become fashionable to talk about the end of the rationale for cities, based on the belief that telecommunications and computer technology will allow each of us to work wherever we want, in a houseboat or in the woods away from cities. A man I sat next to during a flight from San Francisco to Portland, Oregon, epitomized this lifestyle. He ran a software company that was headquartered in Virginia, just outside Washington, D.C., and had its main production facility in Salt Lake City, Utah. He, however, lived in a small town on the Columbia River, near where one of its famous tributaries, the Deschutes River, flows into it. When I asked him how he managed this arrangement, he explained that he commuted by airplane several times each month as needed. Otherwise, he relied on e-mail, telephone, and overnight delivery services. "I want my kids to grow up in the right environment," he said. But apparently he did not share this same concern for the children of his workers; they could endure an existence near the nation's capital. His statements appeared to be those of an elitist who sought a special circumstance that he believed most valuable while employing people who did not have such opportunities.

As his story illustrates, the executive's lifestyle is available to only a few. Think of what the air traffic problems would be, as well as the problems with air pollution, if all the employees of all his factories and all the other high-tech industries maintained this lifestyle. It would be destructive to the environments of city and countryside, destructive to both civilization and nature. Most of humanity needs a different path to a pleasing environment.

Although the technology exists to allow some of us to work in this manner, it is only the well-educated professionals and members of small, elite segments of society who are able to do so. If it were otherwise—if we really could all live like this owner of a software company, then we might decide to abandon our ideas about city life and abandon cities as places to lead our lives. But with present demographic trends, the opposite is the case.

For most people today and in the future, as far as demographers are willing to extrapolate from present data, city life will continue to be a reality and a necessity. Contrary to the idea that cities may become less and less important, trends in human demography show that we are already a heavily urbanized society. In developed countries, 80 percent of the people live in cities, whereas in the poorest countries, only 20 percent live in cities. Trends suggest that urbanization will increase worldwide in the future: as economic development takes place, more and more people will leave the countryside for the city. It is projected that in the future, 50 percent of the world's people will live in cities.

Not only is the human population becoming increasingly urbanized, but also there is an increase in the number of megacities. In 1950, there were just two cities that, along with their suburbs, had populations of 10 million. In 1975, there were seven such cities. In the year 2000, there were twenty-seven cities with more than 10 million people—the population of twenty-five of these urban areas totaled nearly 400 million!

Here is another intriguing fact about the urban momentum. In the future, most people will live in the largest city in any given country. This means that the world will be increasingly urbanized; therefore, if we are interested in helping people live in better environments, we must focus on urban environments. There is no denying this urban trend, and there is no short-term way to deflect its trajectory. Like it or not, we are stuck with cities.

Once these points about population growth are laid out, they may seem rather obvious, perhaps pedestrian, but in my experience they are generally ignored when people talk about what they like in the environment. There are many laws and regulations today that affect urban environments, but most focus on the negatives—preventing air and water pollution. I believe that this emphasis reflects a prevailing negative attitude about cities.

On the other hand, some environmental leaders are beginning to speak of the benefits of cities to civilization and the environment. Roderick Nash, a well-known environmental historian, wrote recently that he hopes that in the future there will be an "urban implosion"—a return to urban life. This, he argued, would have dual benefits, improving the lives of most people while helping to conserve rural land and wilderness. He wrote that his dream for the new millennium is 1.5 billion human beings living in 500 concentrated habitats, creating an island civilization, a renewed urban-dominated civilization. This is an important idea. It suggests that those who are interested in wilderness, as Nash is, and those who are interested in urban environments have a common goal—as we improve urban environments, more people will be able to find recreation there, and there will be less pressure on wilderness. But this is not the prevailing fashion. We continue to admire nature vicariously or briefly, on vacation, and to live out our professional, workaday, and family lives in the suburbs and the cities, disconnected from what we think of as nature.

San Francisco and Portland, Oregon, stand out as highly livable cities in the United States: San Francisco because of the environment in which it is situated, Portland because of what its residents have done to make it a beautiful and livable city. We need to look to cities such as Portland as a guide to our future.

Some suggest that the decline of cities is solely a problem of decaying infrastructure and, therefore, of economics. Recently, I spoke at a Science Day gathering in Washington, D.C., discussing the need for a change in the public's attitude about urban environments. Afterward, a questioner said he believed that people are leaving cities because they are unlivable and the cost to repair the infrastructure—sewers, for example—is so great as to be beyond our society's capabilities. He argued that we must improve the infrastructure first and then people will begin to think about cities in a positive light. I argued that we will not begin to exert the political will and economic power to improve our cities until we once again begin to think of cities as important to civilization and as pleasing environments in which to live. Only when those with money and political power want to live in cities because the cities are beautiful and lively will cities improve as environments.

We must seek ways to bring the feeling of wildness, in Thoreau's terms, into the cities and to bring nature—for physical recreation, for creative inspiration, and, yes, for the conservation of biological diversity—into cities. If we do not do this, cities will fail, and the surrounding countryside will suffer from overuse by those seeking to escape cities and from suburbanization of the landscape.

Once again, a central idea of Thoreau's returns. Although there are benefits in preserving vast areas of wilderness, spiritual contact with nature can be obtained, as Thoreau argued and discovered for himself, in a partially settled countryside, the swamp on the edge of town: a place where one can go on daily walks. This is nature near to or within a city, if the city has been designed to make it so.

Thoreau's famous statement "In wildness is the preservation of the world" may appear to be a rejection of civilization and therefore of cities, but this is not what he meant; quite the contrary. In the sentences that follow this statement, he explained what he meant by the importance of wilderness: "Every tree sends its fibres forth in search of the Wild. The cities import it at any price. Men plow and sail for it. From the forest and wilderness come the tonics and barks which brace mankind." The importance of wilderness is in the way it enriches people and civilization. In Thoreau's metaphorical statement that wilderness provides "tonics and barks"—literally, medicines that improve our physical health—the emphasis is more spiritual than physical, as exemplified by several other statements Thoreau made.

"The African hunter Cumming tells us that the skin of the eland, as well as that of most other antelopes just killed, emits the most delicious perfume of trees and grass," Thoreau wrote. "I would have every man so much like a wild antelope, so much a part and parcel of nature, that his very person should thus sweetly advertise our senses of his presence, and remind us of those parts of nature which he most haunts." There is an essence of wilderness essential to a human being; it is something we can absorb, ingest, as the eland does grasses. Nature is then immersed in us, not us in nature.

Throughout his life, Thoreau struggled with the role of physical wilderness and physical human settlements, his feelings moving sometimes in one direction and sometimes in another. In *Walden,* he commented:

> Our village life would stagnate if it were not for the unexplored forests and meadows which surround it. We need the tonic of wildness,—to wade sometimes in marshes where the bittern and the meadow-hen lurk, and hear the booming of the snipe; to smell the whispering sedge where only some wilder and more solitary fowl builds her nest, and the mink crawls with its belly close to the ground.

But his general attitude was perhaps best stated at the end of his second trip to the Maine woods, when he wrote that it was a "relief" to get back to civilization. What he preferred, most of the time, was something in between a city and wilderness. For him, the best landscapes combined the fertility of the forests with human artifice.

Many of Thoreau's writings evoke wilderness as a background or foundation, metaphorically speaking, for civilization, providing a kind of spiritual fertility, analogous to the chemical fertility of the forest soil. The latter stimulated the growth of vegetation; the former, growth of the human spirit and human creativity. "Even the oldest villages are indebted to the border of wild wood which surround them, more than to the gardens of men," he wrote. Again, it was the juxtaposition of nature and town that was most beneficial: "There is something indescribably inspiriting and beautiful in the aspect of the forest skirting and occasionally jutting into the midst of new towns. Our lives need the relief of such a background, where the pine flourishes and the jay still screams." Here is an appreciation of the city and the forest juxtaposed, intermingled, conjoined. Living what was basically a suburban life with vacations in the country, Thoreau was in this way ahead of his time, perhaps ahead of ours.

Nature is important to the city, and the juxtaposition of city and countryside is essential to what is best about cities—the creativity that city life inspires. Again, Thoreau saw the value of wildness in its influence on human creativity, civilization, and culture, not in preservation of species or ecosystems for themselves.

Given the modern urban momentum and the importance of our immediate environment to our spiritual sense of place in nature, we must confront, and resolve, our ideas about the relationship between urban life and nature. Although many who are interested in the environment per se may not make this connection, people in business have long recognized the economic value of beauty and of vegetation in cities. In 1885, during an early real estate boom in Los Angeles, a group of businessmen tried to develop a place in the eastern part of the Los Angeles Basin—in Mojave Desert country—that they called Widneyville-by-the-Desert. Even today, this is pretty barren country for most people. The eastern edge of the Los Angeles Basin gets little rainfall. Typical of most deserts in the United States, the countryside is one of scattered shrubs and low, tree-like cacti and agaves, with little or no grass and no flowers except during the rainy season. It is a countryside of dry air and bright light. When one adjusts to it, it has its own romance and appeal, an appeal of space and light, of direct confrontation with huge geological structures and formations—tall, stark mountains that have shed their boulders down their steep slopes, their lack of grassy or forested cover revealing their erosion for all to see: naked geological nature. Personally, I love this desert scenery and its dry, brittle, sagebrush-scented air. But for someone brought up in western Europe or on the eastern seaboard of the United States, the first impression of this landscape might be of a horrible barrenness, a hell on Earth; an after-the-bomb countryside inhospitable to grass and, therefore, to human beings; a countryside of dead tans and browns. To the uninitiated, the Joshua trees, with their few scraggly branches sticking out horizontally from the main stem and then turning up at sharp angles like crucified arms, can resemble sculptures of starved and tortured beings. The trees seem to suggest an almost biblical landscape, a wilderness through which one might wander in search of salvation, punished by the environment for one's sins.

At first, oblivious to this, the Widneyville developers brought in people from New York City and other eastern cities and tried to interest them in buying land. But these people, horrified by the sight of yuccas, cacti, and Joshua trees, fled. Unwilling to give up on their investment, the developers bought a trainload of oranges and stuck the bright fruit on the spikes of the yuccas and the cacti, and trimmed the Joshua trees to make them look more presentable. Then they told the next batch of easterners that this was the "natural home of the orange."

If you look at a map of southern California, you will not find Widneyville-by-the-Desert. The development did not succeed. The lesson of this story is that people have long realized the power of vegetation on the landscape and the importance of landscape beauty in an urban environment. It is a lesson that we forget at our own peril and to the detriment of civilization.

I was confronted with conflicting attitudes about cities and nature during a long drought in southern California during the late 1980s and early 1990s. A representative of the Metropolitan Water District of Southern California called me and said that the agency was facing a dilemma. Leaders of environmental organizations and staff members of government agencies in northern California were telling the water district that Los

Angeles should have no trees or lawns because these used too much water. He said that the agency needed to understand the role of vegetation in cities and asked whether I could help. I told him that the best approach would be to combine modern findings from ecological science about vegetation and biological conservation with the great ideas regarding city and park planning of Frederick Law Olmsted, the father of landscape planning in the United States. I contacted Charles Beveridge, editor of the Olmsted papers, and he and I wrote a report about the role of vegetation in cities such as Los Angeles—cities in a semi-arid environment, cities that might arise from the dried oranges of Widneyville-by-the-Desert. Such a city is a severe test of the connection between nature and civilization because it is in an environment in which a high density of people cannot survive without an artificial supply of water—more than that, an artificial supply of everything. Therefore, the conclusion one reaches about the use of external natural resources to support a city such as Los Angeles has broad implications for nature and cities everywhere.

The proposal to eliminate trees and lawns assumes that irrigation of these plants is a significant drain on the water supply for the entire state. In fact, California's cities account for only about 15 percent of the state's water use. Much of the water used in California is for agriculture, especially the production of cotton, rice, alfalfa, and other crops that have a heavy water demand. Most of these are grown in the great Central Valley, itself a desert unless artificially irrigated. The difference between the assumption and the facts seems to illustrate the inherent modern prejudice against cities and their environments; without examining the facts, one assumes that it is the city, not the countryside, that is the great drain on water and the great environmental negative. People approach the idea of cities with, as Thoreau would have put it, "expectation," preconceived notions unsupported by facts.

Another assumption, expressed explicitly in some op-ed pieces and letters to editors during the drought, is that no vegetation is native to southern California—that this was originally a completely barren place, an empty naturalness ruined by the introduction of exotic plant species. This is another false assumption. Los Angeles lies within a Mediterranean climate zone, with mild temperatures, relatively small seasonal variations in temperature, and winter rains that yield a modest amount of precipitation.

This kind of climate is rare, occurring on less than 2 percent of the Earth's surface, but this 2 percent includes some of the most famous and favored places to live in the world: the coast of Greece, southern France, eastern Spain, and most of Italy. It also includes less well known places along the southern coast of Chile, on the southwestern coast of South Africa at Capetown, and in southwestern and southern Australia. These are areas of bright sunshine where freshwater is almost always limited.

In *Two Years before the Mast,* Richard Henry Dana described the Mediterranean climate as it appeared to him in the late 1830s in Santa Barbara, ninety miles up the coast from modern Los Angeles, when California was still a part of Mexico and modern technological civilization had had comparatively little effect on the environment. Today a city of 80,000 and a metropolitan region of about 160,000, Santa Barbara in the late 1830s had about 100 houses as well as its famous mission and a presidio—its fort. "Day after day the sun shone clear and bright upon the wide bay and the red roofs of the houses," Dana wrote, "everything being as still as death, the people hardly seeming to earn their sunlight. Daylight was thrown away upon them."

Born in 1815, two years before Thoreau, Dana was another Massachusetts native who, like Thoreau, attended Harvard University and then traveled. But Dana traveled for his health, not as part of a lifelong quest to understand nature. He suffered from weakening eyesight, which his physician attributed to a case of measles. He urged him, as a cure, to take time off from his college studies and travel. Dana worked on the ship *Pilgrim* that traveled from Boston around South America to California, spending most of its time on the California coast, trading in cattle hides. From this experience came his one and only book, the famous *Two Years before the Mast,* published in 1840, when Thoreau was twenty-three. While Dana was exploring the wilderness of the Pacific Ocean and the sparsely settled coast of California as a sailor on a tall ship, experiencing the real hardships of a sailor's life, Thoreau was making his adolescent voyage on the well-settled and civilized Concord and Merrimack Rivers in the boat he and his brother had built. Dana's book was an instant success—as was Edwin Bryant's book, *What I Saw in California.* Unlike Bryant's book, Dana's has become a classic, read by many a schoolchild. After one other long voyage, Dana spent his life as a lawyer.

Dana's description of the countryside at Santa Barbara is useful to a consideration of the artificial addition of irrigated vegetation, especially of trees and lawns for Los Angeles, because the two cities are in the same climate zone and at that time Santa Barbara's countryside would have had a similar appearance to that of the Los Angeles Basin. Dana described the town as lying "on a low plain, but little above the level of the sea, covered with grass, though entirely without trees, and surrounded on three sides by an amphitheater of mountains." A hill to the east of the town was "high, bold, and well wooded," he wrote. "The town is finely situated, with a bay in front," and the hills behind. "The only thing which diminishes its beauty is that the hills have no large trees upon them," he wrote, "they having been all burned by a great fire which swept them off about a dozen years ago, and they had not yet grown again."

Dana also went ashore in San Diego, a city south of Los Angeles but sharing the same climate zone. When his ship stopped in San Diego, the lack of trees in the region forced the sailors on shore duty to walk long distances to cut firewood. "Wood is very scarce in the vicinity of San Diego, there being no trees of any size for miles," he wrote. "In the town, the inhabitants burn the small wood which grows in thickets," but the sailors "were obliged to go off a mile or two." The trees were mainly tall chaparral shrubs. "These trees are seldom more than five or six feet high, and the highest that I ever saw in these expeditions could not have been more than twelve, so that, with lopping off the branches and clearing away the underwood, we had a good deal of cutting to do for a very little wood," Dana wrote.

Wildlife was sparse. The sailors saw coyotes on these wood-foraging trips, as well as rabbits, rattlesnakes, and crows, and Dana reported that bears and wolves were numerous on the "upper parts of the coast, and in the interior." Dana described a sparse, dry coastline with striking sunlight and mountains but little green vegetation to freshen the landscape.

He walked through coastal grasslands and shrubby chaparral where the only real trees were along streams. It is tough going to walk through chaparral. This kind of vegetation is found in Mediterranean climates. The shrubs that survive are tough, both as obstacles to a hiker and in their ability to survive for long periods without rain. Long droughts make wildfires likely, and many of the plants are adapted to survive frequent fires or have seeds that sprout soon after a fire.

Although they cover only a small area of the Earth, Mediterranean habitats are biologically diverse and support many rare species. Southern California vegetation includes more than 600 plant species, many of which are native. Today, ninety-two of these native species are listed by the California Native Plant Society as rare, threatened, endangered, or of limited distribution.

Few people find this Mediterranean climate and its tough, shrubby vegetation especially beautiful. Those who are unfamiliar with it often find its summer browns and grays oppressive. Few of the shrubs produce brightly colored flowers, and these do so only at certain times of the year. Although the landscape can be a striking bright green after good winter rains, and the upper grassy hills are sometimes colored a brilliant orange by California poppies and a bright purple by lupines in the short spring, for most of the year the landscape appears harsh and uninviting. This raises the question, How does one bring nature that people actually enjoy to the cities of such a region?

Some answers to this question were suggested by the father of landscape architecture and landscape planning, Frederick Law Olmsted, who is best known as the designer of Central Park in New York City. Olmsted traveled widely and developed parks in many parts of the United States, and he had valuable things to say about city planning in southern California. His ideas are intriguing, all the more so because he was a contemporary of Thoreau and had some professional contact with him, and they were fellow New Englanders with somewhat similar backgrounds.

Olmsted was born in Hartford, Connecticut, on April 26, 1822, five years after Thoreau. Like Thoreau, Olmsted was deeply concerned with both nature and civilization. Also like Thoreau, he led an iconoclastic life. From his father, he learned a deep appreciation of the beauty of nature. While Thoreau was writing about the importance of wild nature to cities, Olmsted was seeking to incorporate a naturalistic aesthetic into the design of cities. And while Thoreau tried his hand at amateur farming during his summers at Walden Pond, Olmsted owned farms and was a professional farmer early in his adult life. One of these farms was on Staten Island, now part of urban New York City.

The two not only had professional interests in common but also had professional contact. After his stint as a farmer, Olmsted became editor of *Putnam's Monthly* magazine in New York and published Thoreau's *Cape Cod*.

Olmsted became superintendent of Central Park, then a new park, in 1857, when Thoreau was living in Concord, keeping scientific records about the seasonal changes in nature, and writing his last essays. The following year, Olmsted and his partner, Calvert Vaux, won the design competition for Central Park. This set a new direction for Olmsted's career as America's first and greatest landscape architect.

When Olmsted completed Central Park, he was so pleased about the benefits the park provided for all the people of the city that he went to lower Manhattan, where the poor immigrants lived, and distributed handbills telling them about the new park and inviting them to use it. Olmsted wrote that vegetation in cities plays social, medical, and psychological roles. This is an expression of an approach to life that shares much with Thoreau's.

Today, we tend to believe that the way to commune with nature is to get into the midst of a vast wilderness. Thoreau, having tried that, found instead that he could commune best with nature at some intermediate location—in a physical sense—a swamp on the edge of town. Olmsted, a product of the same era with many of the same interests and concerns as Thoreau, including the broadest concerns with religion, society, and nature, carried these ideas a step further. He integrated the goal of con-

necting people with nature into a practice of designing parks within cities. Thoreau was the thinker and pithy writer; Olmsted, never very good at expressing his brilliant ideas succinctly or eloquently, was the performer, the executor of the idea.

In contrast to Olmsted and land developers of the late nineteenth century, it became fashionable in the modern environmental movement, beginning in the 1960s and 1970s, to consider everything about cities bad and everything about wilderness good. Cities are polluted, lacking in wildlife, dirty, artificial, and therefore bad. Wilderness is unpolluted, full of wildlife and native plants, and therefore good. Ironically, although it has become fashionable to disdain cities, most people, including most ecologists, live in urban or suburban environments and suffer directly from their decline. Thoreau, who wrote, "God does not sympathize with the popular movements," most likely would have looked askance at these sorts of value judgments. In the science of ecology, scientists and practitioners have generally shown little interest in urban environments and have done little research there, although this is rapidly changing.

Today's popular bias against cities is not merely contrary to the ideas of Thoreau and Olmsted; it is contrary to the entire history and philosophy of city planning in Western civilization. City planning has been written about and carried out for more than two thousand years. Throughout this tradition and until our era, those who have written about cities have agreed on three points. The first is that cities have always been civilization's centers of innovation and creativity. The second is that the more pleasant a city is, the more likely it is that its residents will be innovative and creative. The third is that vegetation is the key to making cities livable and is a stimulant to innovation and creativity.

In the history of city planning, there have been two dominant goals: cities have been planned primarily for defense and for beauty. Roman cities were typically designed along simple geometric patterns, which were believed to have both practical and aesthetic benefits. After the fall of the Roman Empire, the earliest planned European cities were walled fortresses designed primarily for defense. Even in these cities, planners considered the aesthetics of the town. For example, in the fifteenth century, planner Leon Alberti argued that large, important towns should have broad, straight streets, whereas smaller, less fortified towns should have winding streets to increase their beauty. Beginning in the sixteenth century, cities began to expand beyond their fortified walls; areas were planted with rows or groves of trees to provide the upper classes with places for promenading and games. In the seventeenth century, Paris began to decorate its boulevards with trees, a practice that soon spread to other cities. By the nineteenth century, street trees were common.

The twin problems of rapid population growth and overcrowding that occurred in Europe's great cities during the industrial revolution resulted in creation of the modern city park, designed to make public open space available to all classes.

The primary purpose of the park movement in the nineteenth century was neither aesthetics nor environmentalism. It was part of a series of sanitary reforms to counteract the threat to health produced by the working and living conditions brought about by industrialization and rapid urbanization. The development of urban open space therefore was not associated with the environmental rhetoric that it is today. Olmsted was part of this movement, and his first goal was not aesthetics per se; he was interested in public institutions that met urban psychological and social needs. Aesthetics were a means to an end.

Regarding the planting of large areas with grasses, Olmsted believed that every large city should have a public park devoted to landscape scenery. The most important element of that scenery was broad expanses of "greensward"—gently rolling lawns and meadows dotted with wide-spreading deciduous trees. With the goal of meeting urban psychological and social needs, he designed park scenery to provide the most effective possible relief from the noisy, fast-paced, artificial, hard-surfaced, and closely built character of the city. One might say that the appreciation of nature he had developed in his childhood, from trips with his father, extended to a belief that such scenery was essential to human health and well-being and to a well-functioning urban society. The park provided a peaceful setting where one could ramble through open space and find relief from the tension and stress of city life. Olmsted even designed certain sections of parks to provide medical benefits for small children and persons convalescing from sickness.

If one were to approach city planning through the logical extension of the extreme ideology of some late-twentieth-century environmentalists—that the only good environment is one that develops without human interference, that it is wrong for people to modify and adjust the environment—

one would either abandon cities as evil or create open spaces within cities where nature is left on its own. Nature would not be assisted in recovering its former complement of species, its structure, and its chemical composition, nor would exotic species be prevented from dominating a site.

If one believes, somewhat more moderately, in the all-encompassing importance of what was natural before human influence—that pretechnological nature was the only really good one and that given the threats to biological diversity everywhere, one is obliged to restore the original landscape in any area wherever possible—then the approach to city planning would be to use all parks and open spaces as opportunities to restore the ecosystems and all constituent species that once existed there. The primary function of parks in a city would be biological conservation and restoration of natural ecosystems. No exotic species would be allowed. With this as a goal, even urban residents would be only occasional visitors to the parks–nature preserves. Recreation might be restricted to viewing restoration projects or participating as volunteers in these projects.

Olmsted's approach was not so much to let nature reign, in the sense of putting areas within a city aside and letting whatever happened happen, as to carefully construct naturalistic settings, to use artifice to imitate natural scenery. Through careful shaping of the land and construction of well-drained and well-surfaced all-weather walks and roads, followed by creation of open space integrating grass, shrubs, and trees, Olmsted used skills in both engineering and landscape design to create particular, restorative environments, including areas of "wildness" such as the Ramble in New York City's Central Park. Technology was put to use to help people and civilization.

For Olmsted, a park's social purpose was as important as its psychological and public health roles. The large landscape park was to be the one place where all elements of a city's population could gather and mingle without the competitiveness and hostilities of the workaday world. In some cases, they would join in common amusements—public gatherings or concerts—but much of the time they would simply walk, picnic, and play with family and friends, conscious of shared pleasure in a place owned in common with their fellow citizens. The park was to be the yard for those without yards, the destination for persons unable to frequent popular vacation spots, and the central social and gathering place of the city.

Although broad expanses of greensward were an essential part of Olmsted's designs for parks in the East and the Midwest, he believed that such landscape features, based on styles that had developed in the rainy climate of England, were inappropriate for the mountain states and California. The semi-arid West, he concluded, called for development of a new style of landscape design. For one thing, the cost of irrigating large areas in order to secure good turf was too high. In addition, he hoped to design a landscape based more directly on the region's natural scenery and climate.

Olmsted made this clear in the advice he offered to William Hammond Hall during the design of San Francisco's Golden Gate Park in the 1870s. The park's shape and size were closely patterned on those of Central Park. Olmsted objected to this approach and to the idea of replicating eastern landscape practices in California: "I have given the matter of pleasure grounds for San Francisco some consideration and fully realize the difficulties of your undertaking. Indeed I may say that I do not believe it practicable to meet the natural but senseless demand of unreflecting people bred in the Atlantic states and the North of Europe for what is technically termed a park under the climatic conditions of San Francisco."

Instead, Olmsted urged Hall to experiment with plant materials that would make possible broad landscape effects of "rich, constant and varied verdure" with little watering. He wrote:

> Cutting yourself completely clear of the traditions of Europe and the East, and shaping your course in details by no rigorously predetermined design, but as you find from year to year that nature is leading you on, you will, I feel sure, be able to give San Francisco a pleasure ground adapted to the peculiar wants of her people, with a scenery as unusual in parks as the conditions social, climatic, and of the soil, to which your design is required to be accommodated.

Olmsted's central idea was that each region has its characteristics and its kind of nature. There is not just one nature that can be meaningful to people, that can meet their physical and spiritual needs. A city needs to be designed within the geology, climate, and vegetation of its location.

The idea of irrigated lawn as, properly, a communal possession was repeated in two of Olmsted's proposals. One was for the campus of the

College of California and an adjacent residential neighborhood, which Olmsted planned at the same time and published in the same year, 1866, as his proposal for San Francisco's pleasure grounds. He proposed to group the college buildings on high land on the property and use the surrounding areas for residential lots. In a lower section, he proposed to set aside twenty-seven acres for a park, with spreading shade trees around its edges and in the center "a perfect living greensward."

In their design suggestions for southern California, Olmsted and his son, Frederick Law Olmsted Jr., rejected the extremes: either a city without vegetation or one so heavily irrigated as to resemble a city in England. Based on an analysis of landscape planning in semi-arid climates, Olmsted proposed four principles of semi-arid landscape design: first, to leave little bare ground exposed to view, instead planting vegetation wherever possible; second, to arrange heavily visited places so that vegetation in and near them can be easily watered, keeping dust and dryness to a minimum; third, to plant vegetation so that it frames distant vistas and obscures the dusty middle distance common in semi-arid environments; and fourth, to plant as much vegetation as possible around buildings so as to connect them visually to the surrounding countryside.

Olmsted and other landscape planners of the nineteenth century established a basic philosophy and many design attributes of city planning that are still much in use today. His approach is in keeping with what I have described as Thoreau's and is in the tradition of Western civilization's rationality. What we need to do is integrate this philosophy and these basic design attributes with modern environmental needs and the new understanding of ecological systems.

The primary difference between Olmsted's approach and a modern environmental approach is the concern with biological conservation. Olmsted saw the use of vegetation as a design problem and viewed vegetation as a palette with which a landscape architect painted a cityscape. He thought that the native vegetation of southern California provided too limited a palette and chose to include plants from other regions that would thrive in that region's microclimate. If the primary goal were biological conservation, aspects of this scenic approach might have to be sacrificed.

The basic structure of a city as designed by Olmsted can benefit biological conservation by conserving riparian zones and wetlands and creating transportation corridors that can also accommodate the migration of animals and seeds of plants. Olmsted believed that a city should have a continuous green thread of parkway. Part of the parkway's purpose was to promote efficient movement through the city, particularly from the city center to outlying parks and from one part of the park system to another. Today, we can design or modify such greenways to become ecological corridors.

This use of urban and suburban river corridors is also consistent with Thoreau's daily walks near Concord. During the last ten years of his life, Thoreau often strolled alongside rivers and watched wildlife. On Christmas Day 1858, he noted in his journal: "I see, in the thin snow along by the button-bushes and willows just this side of the Hubbard bridge, a new track to me, looking even somewhat as if made by a row of large rain-drops, but it is the track of some small animal. The separate tracks are at most five-eighths of an inch in diameter, nearly round, and one and three quarters to two inches apart, varying perhaps half an inch from a straight line." He drew a picture of them. "Goodwin, to whom I described it, did not know what it could be."

On another winter day, December 22, 1852, Thoreau wrote: "You cannot go out so early but you will find the track of some wild creature. Returning home, just after the sun had sunk below the horizon, I saw from the Barrett's a fire made by logs on the ice near the Red Bridge, which looked like a bright reflection of a setting sun from the water under the bridge, so clear, so little wind, in the winter air." Here, he appreciated wildlife and found a neighbor's fire part of the beauty of the local landscape.

In his chapter on economy in *Walden,* Thoreau related: "For many years I was self-appointed inspector of snow storms and rain storms, and did my duty faithfully; surveyor, if not of highways, then of forest paths and all across-lot routes, keeping them open, and ravines bridged and passable at all seasons, where the public heel had testified to their utility." We can take this as an expression of Thoreau's appreciation of paths in the woods near a city. He was walking through a heavily modified but naturalistic and pleasing landscape, much like an urban greenway envisioned by Olmsted.

In our day, urban corridors, greenways, and parks, as well as wetlands, riparian zones, and other areas representative of native ecosystems within urban areas, can be places for recreation, spiritual uplift, and creative inspiration. They can also aid in conservation of biological diversity and protection of endangered species.

To return to the argument that there should be no trees or lawns in Los Angeles, one problem is that this represents a single-factor approach, in which one factor is taken into account at a time. For example, the desire to cut back on water use is not linked with a simultaneous desire to conserve and restore wetlands within the city of Los Angeles. It is popular today to say that nature is complex, but in practice we rarely take that complexity into consideration. Thoreau, in contrast, recognized the need for a variety of approaches to a landscape.

Thoreau, Olmsted, and their contemporaries saw nature and civilization as intimately linked; in contrast, many people today, both environmentalists and their opponents, see them as opposed. They saw nature as functioning to benefit civilization and cities. They believed that cities had essential roles for people. If we are to save nature in the future, we must return to aspects of Thoreau's and Olmsted's philosophy. As urban momentum carries us forward to an ever more urban world, wilderness, endangered species, and civilization all require that we make cities livable once again, for people and for nature.

Civilization and Nature

*"**Civ • i • li • za • tion** . . . n (1772) **1 a** : a relatively high level of cultural and technological development; specif : the stage of cultural development at which writing and the keeping of written records is attained."*

Merriam-Webster's Collegiate Dictionary, 10th ed.

"Two men in a skiff, whom we passed hereabouts, floating buoyantly amid the reflections of the trees, like a feather in mid air, or a leaf which is wafted gently from its twig to the water without turning over, seemed as in their elements, and to have very delicately availed themselves of the natural laws. Their floating there was a beautiful and successful experiment in natural philosophy, and it served to ennoble in our eyes the art of navigation, for as birds fly and fishes swim, so these men sailed."

Henry David Thoreau, *A Week on the Concord and Merrimack Rivers*

DURING THE last ten years of his life, Henry David Thoreau formalized his observations of nature in his daily walks near Concord. He visited the same plants over and over, in different seasons, in different weather. He recorded the days on which trees, shrubs, and flowers leafed out, flowered, bore fruit, dropped fruit, when their leaves turned color and when they fell. He made lengthy and detailed daily entries in a journal. On October 24, 1853, he wrote, "Black willows bare" and "Most alders by river bare except at top," and then, "The rock maple leaves a clear yellow: now and then [one] shows some blood in its veins, and blushes."

Next, he abstracted these individual observations into another journal to summarize major events and seasonal changes. Finally, he began to create tables of his observations, listing species in the left-hand column and years across the top. He was working on these tables at the time of his death. The tables record changes that occurred with the seasons, such as the date on which leaves of a certain tree species fell.

One table, titled "Fall of the Leaf," contains partially completed entries for white pine. For 1852, he wrote in the row for white pine, "Oct 23 have shed"; for 1854, "October 16 fallen?"; for 1856, "October 22 pine needles falls"; for 1857, "November 8 ground in swamp sometimes covered with needles"; for 1858, "October 22 mostly fallen." Botanists call these phenological records. With such tables, the differences among years are explicit and clear. Thoreau's are among the most detailed repeated records ever made of seasonal changes in nature. Ralph Waldo Emerson referred to them as Thoreau's great unfinished work.

Each level of Thoreau's records served a different but complementary purpose. The tables were scientific information; the daily journals were a record of his physical and spiritual contact with nature—the condition of nature and its effect on Thoreau's emotions, moods, and ideas.

Whereas Thoreau's tables were summaries of physical events over the years, his daily journals were more. They were a combination of impressions, reactions, and emotional responses with careful, detailed observations of plants and the seasons. "It has rained all day, filling the streams," he wrote on October 24, 1853. "Just after dark, high southerly winds arise, but very warm, blowing the rain against the windows and roof and shaking the house. It is very dark withal, so that I can hardly find my way to a neighbor's." He expanded into metaphor: "We think of vessels on the coast, and shipwrecks."

The approach Thoreau used during the last ten years until his death can help us, in this new millennium, make a major transition in our thinking about nature, about the role of people and civilization in relation to nature, and about the effect of nature on people and civilization.

We are living through a time of great transition in our ideas, beliefs, and feelings about these fundamental aspects of the human situation. On the surface, this transition appears as a set of questions limited to physical phenomena, questions about what we call "environmental problems." But more fundamental are the attitudes we retain toward the intangible qualities of nature and the connection between human beings and nature.

During the latter half of the twentieth century, a major change took place in the scientific understanding of nature on Earth. Ecology, geology,

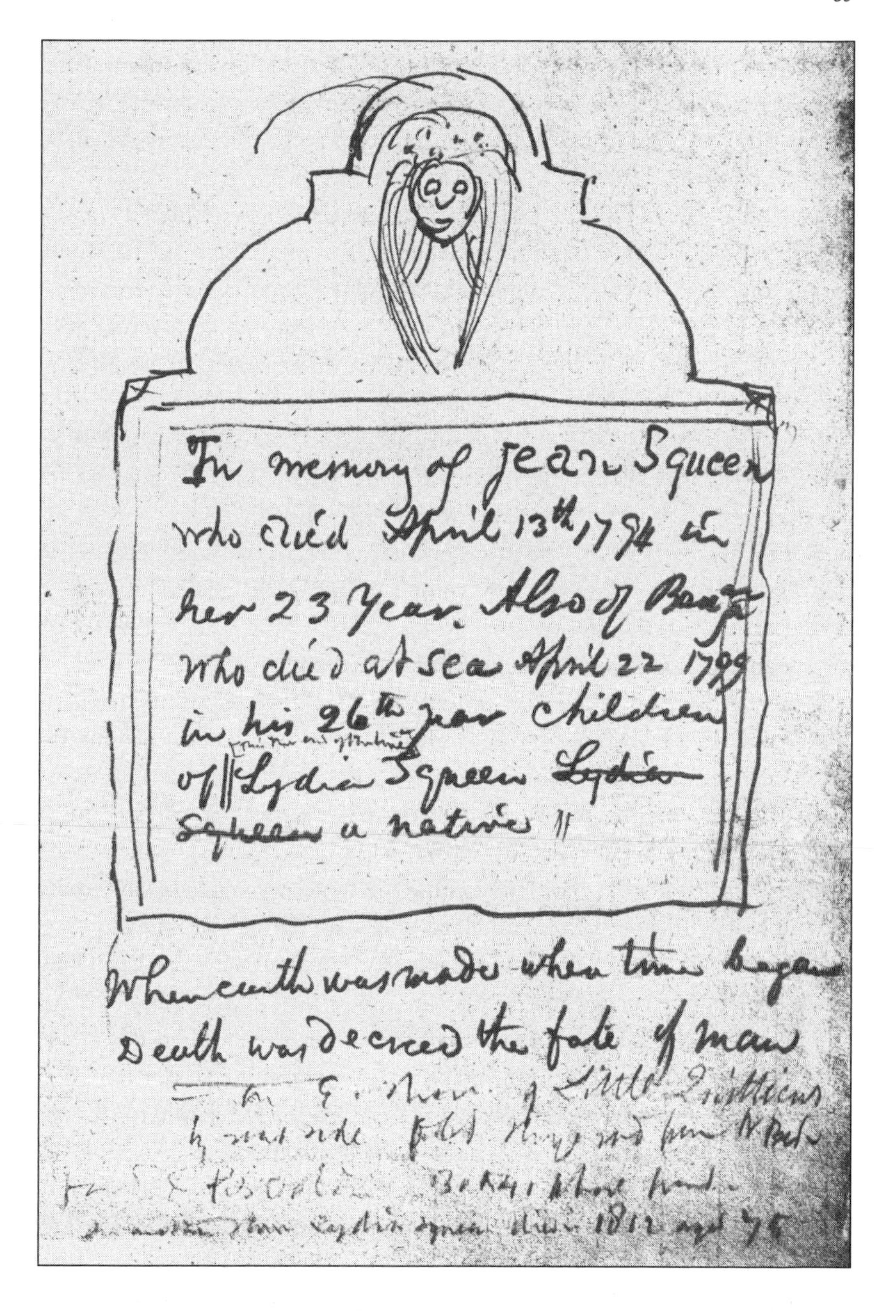

hydrology, geography, oceanography, and paleontology, among other sciences, produced findings that contradicted established beliefs. When the understanding of nature's physical characteristics undergoes a major change, this leads to corresponding changes in the way we perceive our spiritual and religious connection with nature. These are very difficult changes to experience. We tend to sublimate them; if we do recognize them, we resist them. Our established beliefs are so deep-seated that they function as unstated premises of our lives, and we may not recognize that the basis for these beliefs, in our observations through our senses and technological extensions of these senses, has changed. Yet we can come to terms with changes in our nature-knowledge only when we face them directly, in all their manifestations.

This is a two-way process. Our concept of nature affects and restricts our observations, and our observations change our concept of nature. When we resist a change in our beliefs about the intangible qualities of nature, we end up also resisting what the new observations tell us about nature. In this way our beliefs interfere with our knowledge.

An analogous transition took place with the rise of physical science at the beginning of the scientific-industrial revolution. But the transition was only partial and is yet to be completed. This is because in Western civilization, three aspects of nature have amazed people and provided a foundation for classical arguments for the existence of God. These are the perfection of the cosmos, the perfection of the human body, and the perfection of biological nature. Perhaps these began with observations of the wonders of nature, the wonders of the stars; the wonder of the adaptations of our own bodies—as with our eyes and our hands—to meet our needs and enable us to alter our surroundings; and the wonder of the wealth of species in the world, each with a different lifestyle and each amazingly well adapted to that style. Life functions and persists within an environment both threatening and nurturing.

It seemed impossible for these three features of the universe—cosmos, body, and biosphere—to have come about on their own, without a maker. They seemed to be evidence for the existence of some higher being or beings capable of creating them. In the Judeo-Christian tradition, belief in the perfection of these three features became necessary because the God of the West was both omniscient and omnipotent and therefore had to have made the world and biological nature perfect. Finding or claiming to find an imperfection in nature could be seen as heretical. A perfect and all-powerful God could not have made an imperfect creation.

This was the problem between Galileo and the Catholic Church. Galileo was the first person to use the telescope in a methodical study of the cosmos. In the early seventeenth century, he observed the movements of the stars and planets and collected evidence that Earth revolved around the sun, that the moon's surface was rough and irregular rather than smooth, that the sun's surface had dark spots, and that Jupiter had its own moons.

Johannes Kepler also contributed to the change in the conception of the cosmos, demonstrating that the planets moved in elliptical rather than circular orbits. A circular orbit was taken to be perfect, and therefore Kepler's findings seemed to be another blow against structural perfection of nature.

One implication of these observations was that Nicolaus Copernicus had been right: Earth revolved around the sun, not the sun around Earth, and therefore Earth was not at the center of the universe. Another was that there seemed to be structural imperfections in what had been seen as a perfectly created, human-centered universe. Resolution of these conflicts came with the development of Sir Isaac Newton's laws of motion. These, expressed elegantly in mathematics, suggested that there was a different, previously unrecognized kind of perfection in the cosmos: a perfection of processes, motion, and dynamics. God could have created a cosmos that was perfect in the laws that governed motion, a perfection at a higher level than that of structural perfection.

In classical Western thought, from the Greeks and Romans to Galileo, symmetry was characteristically seen as a necessary attribute of beauty; mountains were considered unbeautiful because they were asymmetrical objects and therefore imperfections on the symmetrical surface of the Earth. A Roman garden, as represented to some extent today by the classical Italian garden, is considered beautiful because it is symmetrical, but the symmetry has been imposed by people on an unsymmetrical nature. To take this view of landscape beauty to its logical conclusion, our modern earthmoving abilities would lead to an appreciation of, and construction of, artifacts of great symmetry. Of course, we appreciate this kind of beauty when we seen it in architecture—in the Empire State Building, the Golden Gate Bridge, the Eiffel Tower.

To believe that beauty and perfection in biological nature can be found only in a static and completely symmetrical scene is to fall victim to the fallacy of confusing structure and process. It is to fall victim to the belief that the proper state of nature is defined by a static structure rather than understanding that biological nature is an unfolding, ever-changing process.

The conflict in Western religious thought brought about by the change in scientific knowledge of the cosmos was resolved through a recognition that there is beauty and perfection in the laws that govern the dynamics of the cosmos—that Newtonian mechanics represent a kind of perfection. Dynamics replaced stasis. It was a difficult transition. Once this resolution became part of Western culture, structural perfection was no longer considered a necessary aspect of beauty. At the dawn of the nineteenth century, the Romantic poets drew a distinction between the simply beautiful and the sublime, awesome, and grand. Symmetry, as in the arching shape of an elm tree, was pretty. The grand arc formed by Niagara Falls was called sublime because of its scale and the motion of the water. Mountains, ocean storms, and storm waves were considered awe-inspiring, grand, and therefore aesthetically pleasing in their representation of power. From a religious-aesthetic point of view, they were a demonstration of the power of God and the perfection of God's dynamic laws of motion. People began to visit mountains as places of beauty, recreation, inspiration, and creativity. Poets wrote of the power of mountains and ocean storms. Mountain climbing became a form of recreation, and new technologies made such visits possible for a wider and wider audience.

But that first transition covered only the idea of the cosmos; today, we are in the midst of an analogous transition in our view of biological nature. From a historical perspective, this is a completion of the modern transition in our entire relationship with nature. Because we resist the changes in ideas about biological nature and hold on to the "balance of nature" beliefs of the past, our nature-knowledge, as Joseph Campbell called it, is four thousand years old and does not agree with late-twentieth-century findings about biological nature.

Given a belief that perfection lies in static nature, in a single "best" state, the discovery that biological nature is always changing once again poses a threat to religious beliefs and our feelings about our place in nature. In the late twentieth century, much of the response to this perceived threat was to ignore the new knowledge, pretending that it either did not exist or was not relevant beyond the surface of our existence.

Some theologians have told me that these kinds of observations are of no consequence to their religion because they deal merely with the phenomenological, the tangible, which is but a faint shadow of God's powers and meanings. But like it or not, accept it or not, our perceptions of biological nature and of our connection to it do affect our spiritual sense.

Within the context of Western civilization, which has treated religion as a historical process, a theologian who ignores this change risks having his religion become irrelevant to the masses of people when religious precepts no longer fit human needs, no longer answer human questions, no longer are congruent with the emerging worldview.

The resolution of today's conflicts between scientific evidence and existing beliefs and mythology about nature will parallel the earlier ones concerning astronomy and Western religions. Our new nature-knowledge tells us that the living world is dynamic, not static. Ecological systems are so much more complex than the solar system, and the great minds of today have been so little concentrated on the subject of ecology, that we do not yet have an internally consistent, mathematically elegant theory of ecology to parallel the Newtonian laws of motion. From where we stand, it is unclear whether such simple, elegant laws can indeed emerge for such complex systems.

This makes the transition that much more difficult. But the key, once again, is that the beauty in the dynamics of nature can replace the beauty of the idea of stasis. In this case, nature's ability to respond to change and life's demonstrated ability to persist for unimaginably long times—3.5 billion years—are ideas of great aesthetic appeal. Life has created order where

there was disorder—on a small planet in the midst of a vast universe of increasing disorder. Biological nature has the power to recover, to restore, to heal itself. Even more powerful is its ability to evolve to meet new challenges and opportunities. This sunlit island in the black sea of space is a place of great order, a place of biological adaptation and invention. It has its rules; it is neither chaotic, in the classical sense of lacking form and rules, nor rigid. This understanding of nature has its own aesthetics. It has the beauty of a palm tree bending in the winds of a storm, of salmon rising over rocks in a stream on their way home to spawn.

The new observations of nature therefore lead to a new aesthetics, one of beauty in process rather than only in structure. Where is the place of human beings within this dynamic, biological nature? As has always been the case, there are two classical opposing opinions about the role of human beings. To use nineteenth-century mechanistic terminology, one opinion is that human beings are the final cog in the wheels of nature's machinery and that if we function in our role in the great chain of being, nature will be complete. The other view is that biological nature is perfect without human action and therefore human actions are an interference, pulling out cogs in the machinery of nature.

We see the two opinions rephrased in modern environmentalism. The belief that disaster lies ahead for all life on Earth as a result of the seemingly unstoppable growth of the human population, in turn a result of our technology, and as a result of growth in the kinds and power of technology is a modern expression of human beings as an interference with nature's perfection. Garrett Hardin best expressed the latter problem in coining the term *technological imperative.* By this, he means that it is generally accepted today that if we can imagine something, we can make it, and if we *can* make it, we *must* make it. This is an assertion that both the technologies that are good and those that are bad for the persistence and diversity of life will be made and will increase in numbers and power.

The implication is that growth of the human population combined with growth and complexity of technology will overwhelm and destroy the biological world, not just locally but globally, destroying the biosphere and destabilizing the entire planet's life-containing and life-supporting system.

Frustrated with humanity's failure to control its population growth and to select wisely among technologies, some conclude that all technology is bad. They believe that given the human tendency toward moral imperfection, *Homo sapiens* must be considered corrupt and corrupting to both the structure and the processes of nature. The deep ecologists conclude that people must therefore be placed at the bottom of the moral order. Recognizing that all life depends on the functioning of the biosphere, deep ecologists see our primary moral imperative as preservation of the dynamic global life-including and life-sustaining system. Man and woman and all our works—our rationality, creativity, and innovativeness, our art, architecture, and music, our poetry, philosophy, and science—are to be rejected if they interfere with the functioning of the planet's life-sustaining system. The conviction is that this will of necessity happen.

There is no denying the truth in Thomas Robert Malthus's original statements about the potential for human population growth, especially if one takes the time to read them in their entirety rather than grab at them as slogans. A finite world can support only a finite number of people. There must be an upper limit. But there is a trap in this argument if it is the only argument on which one focuses. As an illustration, I was asked to visit a small town in the California desert near the mountains that separate the great Central Valley from Death Valley. The town had an environmental issue that the citizens could not resolve. Developers wanted to build new housing, but a small ground squirrel had been listed as a threatened or endangered species and its distribution was not easy to determine. A developer might buy land and begin planning to build on it only to learn that since completion of the purchase, a squirrel's burrow had been discovered on the land. If this happened, the developer could not proceed and would lose most of the value of the investment. How could this problem be resolved?

I was asked to give a talk and help lead a public meeting. At the meeting, one of the developers said: "Well, isn't it true that the really important environmental issue is human population growth? Shouldn't we be spending our time on that? Why should we waste our effort in conserving this small animal if all of nature is threatened by our own population? Why can't you just ignore this local problem and let us build our houses while those of you who are concerned about the environment go and do something about the real problem?" The environmental argument was turned on itself so that the local was to be sacrificed for the global.

The speaker reminded me that human population growth can be used as an excuse for rejecting any other effort to conserve nature. The lesson of that public meeting was that we cannot wait until the human population problem is solved before we tackle the environmental problems at hand or before we try to find out how nature and civilization might be compatible. If there is to be a future for life on Earth, dynamic ecological systems must function, their processes must be maintained, and some amount of biological diversity must be conserved.

An alternative to the two opposing approaches—the first rejecting human endeavors and the second embracing any human actions except those to control total human population growth—is to understand which processes are "natural" in the sense that living things have been subjected to them for very long times and therefore have adapted to them. With this understanding, we can select those technologies that are likely to be benign and those that are not. This perspective on humanity and nature also allows us to see that many human creations and inventions form a continuity with the historical process of the evolution of life on Earth—that human inventions are part of the dynamic process of evolution. When human beings began to invent and create, nature was changed and the development of art, poetry, music, and science was initiated.

Civilization and human creative impulses are both fragile and powerful. Civilization is fragile in the following way. The greater our technology, the greater our sense that we are independent of nature and independent of the anti-civilized drives deep within us. But the greater our technology, the greater our power over nature and the more dependent we are on nature's resources in the largest sense of the word, and therefore the more vulnerable we are. It is what the great historian and geographer Lewis Mumford said about cities. The artificiality of cities gives city dwellers a sense of being independent of their surroundings, of nature. But the more artificial the city, the more it depends on the surroundings for its resources. And so there is a strange contradiction: the more secure a city dweller feels in his independence of nature, the more he depends on nature and the more vulnerable he is to it. So it is with civilization in general.

Civilization leads to highly visible, tangible products such as cities and rectangular plots of cleared land yet is itself an intangible system contained within the thin skin of life on our planet. It needs to be nurtured. Once started, however, civilization and human creativity are not easy to stop. The more that we see human beings as separated from nature, the harder it is to evaluate which kinds of creativity support the world's biological system and which do not. Perhaps there is a path other than Hardin's technological imperative. If so, its basis must be a new aesthetics, an understanding of the value to people of the intangible qualities of nature and, as Thoreau's experiences suggest, even of a sense of fun. With this perspective, we may be able to see civilization and nature as interconnected systems that are both of value.

I have suggested that there is a path that can lead to the compatibility of civilization and nature. If we take that path, can we achieve this with perfection—are we likely to act in a mechanically perfect fashion? Once the question is posed, the answer seems self-evident: people tend to muddle through, and societies are imperfect. I am reminded of a story in one of my favorite books, a small, obscure book from my father's library titled *English Wayfaring Life in the Middle Ages.* A wayfarer myself, I have found this book appealing. It describes the few groups of people who were wayfarers in the Middle Ages: troops of actors, outlaws, outcasts, traders, those on religious pilgrimages. It also describes how people traveled or "wayfared" during that time. One series of events was repeated over and over. When travelers reached a river and needed a way to cross, they would pay the farmer who lived next to the river to row them to the other side.

Eventually, this business would become the farmer's main activity. He or his sons would improve on the situation by building a toll bridge and collecting money from wayfarers. Often, the original bridge was solidly built and lasted a long time, but over time, with wear and tear and the effects of storms and floods, the bridges would eventually decay, falling apart piece by piece. The process of decay occurred quite slowly at first. There would be some complaints from the wayfarers, but the response would be that the bridge had been there for a long time without falling down, and this was proof enough that there was no danger to the traveler or to the bridge.

Eventually, the gradual process of decay, combined with the farmer's lack of foresight or will to make repairs, persisted until the bridge collapsed. A great complaint would arise, and a new bridge would be built. The process of use and decay would be repeated, warnings about impending disaster would be ignored once again, and the bridge would fall down, only to be followed by another sequence of rebuilding, complaint, and collapse. It seems to have been an endless process in which people never learned from past mistakes, responding only to disaster.

Having told this story often, I was not surprised to read a recent newspaper article announcing that most of the bridges in the United States had been neglected and were in danger of collapsing. In the greatest technological nation the world has ever seen, human abilities to deal with future problems appear to be about the same as in the wayfaring life of the Middle Ages. We muddle through. We avoid dealing with disaster until we have to. This, then, is our likely path concerning the environmental issues we face today—all the more so if we do not make use of Thoreau's approach as described in this book. During the last decade of Thoreau's life, this approach included careful observation of the details of nature near home, like those with which this chapter began. If we were to apply this approach to the maintenance of bridges, we would carry out a detailed monitoring of changes in their integrity, their soundness, and then, like Thoreau the engineer of pencil production, we would do something about them.

It would not be surprising if our ability to deal with the much more abstract idea of a world too full of people were no better than our ability to fix bridges in a timely manner. Given the history of human actions, it seems likely that people will not act in a timely and mechanically rational way to ease the human population down to a hypothetically sustainable level and keep it there. It is much more likely that the human population will undergo busts and booms, as do all the animal populations for which we have data. Before we get around to saving ourselves by controlling the size of our population, our numbers will probably be reduced by new diseases and environmental catastrophes, like the floods in Bangladesh of recent decades. For the people who suffer, this is a terrible future, but it is a likely one.

I have used Thoreau as a metaphor for several ideas that I believe can help us avoid such mistakes, if we have to the will to do so. Thoreau also serves as a metaphor for how we might resolve apparent societal conflicts between those who want to put nature first and those who want to put civilization first. The resolution involves a set of stages of realization. These include recognizing the many reasons to value the conservation of nature, distinguishing clearly between spiritual needs and tangible needs, distinguishing between means and goals, and developing an approach to knowledge about nature that integrates the scientific method, used properly, and the insights of experts, both experiential and professional.

The approach I am suggesting is that of a rationalist, in the traditional meaning of the word in Western civilization. This philosophical perspective accepts the fact that human beings are the only sentient creatures on the Earth and therefore are the only ones who can make conscious choices—that the moral issues remain a human determination, whether we like it or not. We are stuck in the driver's seat, even if we wish we were merely passengers.

To take this approach, we begin, as Thoreau did in determining the depth of Walden Pond, by examining the available facts. While attending a conference several years ago, I was struck by a statement made by Roger Sedjo, an economist who specializes in forestry issues. He began by comparing the amount of timber traded internationally for use in construction and for pulp and paper with the amount of timber that can be grown on highly productive forest plantations. It turned out that the timber needed for world trade in these commodities could be grown on less than 5 percent of the total forested area of the Earth. That is a striking statement. Sedjo and I subsequently discussed this comparison and wrote a longer analysis. The timber used in construction, pulp, and paper accounts for 90 percent of the wood traded internationally; the rest is hardwoods used for furniture and other specialized purposes. The amount traded internationally for construction, pulp, and paper can be taken as an estimate of the total world use of wood for these purposes. It does not account for domestic use of firewood, which is not considered an international commodity.

The next step in the process Thoreau followed, which a modern scientist also would follow, is to convert these percentages into quantities. Each year, 1.5 billion cubic meters of wood are traded for use in construction, pulp, and paper. Reasonably high-yielding, but not the most productive, tree plantations grow about 10 cubic meters annually per hectare (a unit of land area equal to about 2.47 acres). With such productive forestland, the timber required for international trade could be produced on 150 million hectares (about 371 million acres, or about 579,000 square miles), or about 5 percent of the world's existing forestland.

In contrast, typical rates of production from nonplantation forests (forests that regenerate "naturally" and are not managed to increase their production) are 1–2 cubic meters per hectare. If we were to try to produce all the timber traded from this kind of forest, we would need 25 to 50 percent of the world's forestland. Clearly, when a specific use requires 50 percent of the world's forested areas, there will be serious conflicts over different desired uses. But at a level of about 5 percent or less, one can imagine a world in which the timber needed for international trade is produced on such a small fraction of forestland that most forests are left to meet other uses.

This simple calculation would seem to open up many possibilities. We could have a variety of forested wilderness areas, areas for timber production, areas for recreation that did not require a wilderness setting, and areas that could be managed to conserve endangered species or species of other special interest. There would be room for land on which no direct human action is taken and room for a combination of nature and civilization like that preferred by Thoreau. Most forests could be devoted to aesthetic and ecological purposes, designated as parks and preserves.

This calculation moves the issue to a different level from the conflicts of the past, as illustrated by a discussion of these ideas that I had with Chad Oliver, a professor of silviculture at the University of Washington. He suggested that the simple partitioning of forests into nature preserves and intensely productive plantations would not work in practice because it would create a black-and-white world in terms of forests: huge areas with no commercial production and small, intensely managed areas. He suggested that over time, the huge areas, without some economic value, would tend to receive less and less attention in terms of conservation effort. A more practical arrangement, he argued, would be a finer mosaic with commercially productive land interspersed among conservation lands.

Oliver's response to the analysis Roger Sedjo and I had done moved the discussion from a highly charged, emotional one to a rational exploration of facts, processes, strategies, and tactics—from rhetoric with a weak basis in fact to an examination of the implication of facts. The shift was toward a Thoreau-like approach as I have described it.

The new discussion about forests raises a set of questions that can, at least in theory, be answered by scientific analysis, such as determining the geographic pattern of a landscape that is best for forest conservation and also produces timber for use. In this way, the discussion moves to real factors that have to do with timber products. The requirements for analysis and practice become clear.

The idea of using high-yield forest plantations and creating a geographic separation in the production of timber and the conservation of forest ecosystems seems so obvious and simple that it is strange that it is not widely known and discussed. It would seem an obvious way to conserve biological diversity that requires forestland and meet all the other needs for forests that are put forward as environmental functions and goals. If these concepts are so simple, why have we not approached them in this way before?

I believe they have not come to the fore in discussions for four reasons, all of which relate to Thoreau's approach. The first is that no one had made the calculations suggesting that a small percentage of forestland could provide the timber needed for international trade. The second is that the term *forest plantations* is viewed with "expectation"—taken as a bad word by some who are concerned about the environment because it appears to represent an attempt to convert the few remaining old-growth forests into plantations. The actual suggestion is to leave old-growth areas untouched, instead developing plantations where land has already been disturbed, such as where it has been converted to agriculture but is not productive for this use. The fear is that an emphasis on plantations might be a first step down a slippery slope, ending only when all forests were converted to plantations. The third reason why these ideas have been overlooked is that some would consider such an analysis unrealistic, arguing that people are not able to manage anything as complex as forests.

The fourth reason is a concern with the human population problem—that no use of resources can be sustainable as long as the human population continues to expand. This is, of course, true. But even though we must attempt to reduce human population growth, we cannot stand by idly and let the forests of the world be seriously damaged and destroyed.

Can we expect that in the future, such an approach will work perfectly? I doubt it, but I do believe that efforts in this direction will contribute to a better world—better for both civilization and nature—than will the path we are now pursuing. It would be naive to assume that we could put this approach into effect immediately—not least because we lack the data that Thoreau recognized was so important—but we can set a direction that will improve our use of natural resources and benefit civilization over the long term.

The timber production example is just one illustration of how we can take a different, still rationally based approach to compatibility between civilization and nature. To generalize, the first step is to accept that we value nature for two quite different qualities, which I have referred to as the tangible, or physical, and the intangible, or spiritual. We have to confront our desire for both of these. Then we need to ask, What are the characteristics of a nature that meets our intangible, spiritual needs? Thoreau found that his spiritual needs could be met in a partially settled landscape with places one could retreat to, such as a swamp on the edge of town. It did not need to be huge. It did not need to be untouched by human activities. The image that comes to mind is a map of what modern ecologists refer to as a mosaic of different landscape types, interconnected, serving different uses in different places.

The next step in resolving our conflicting ideas about nature and civilization is to recognize that the human spirit is tied up as much in civilization as in nature. If life on Earth exists as a thin skin on the surface, fragile in many ways, on a planet that is unique or at least rare in the universe, then human creativity, including the books, ideas, scientific advances, and spiritual qualities that Thoreau valued, exists in an even more fragile system, the intangible system we call civilization.

Demographically, Thoreau was much like most of us today: he was an urban or suburban resident. He was also like college-educated people of our time in that his outdoor experiences were more suburban than wild, with a few major exceptions—his experience on Mount Katahdin being perhaps the major outstanding counterexample. He went into the wilderness on holiday trips, leaving his home in Concord, Massachusetts, and traveling to the forests of Maine. A short vacation trip is the way most of us experience wilderness if we do so at all.

Thoreau also was a product of a technological society and made technological contributions to it. His path to a relationship between himself and nature, and between civilization and nature, would seem to arise from much that is familiar to us today and could be a comfortable approach for us as well. But we will need to make some changes if we decide to follow Thoreau's path.

One such necessary change is at the personal level—we will need to find some kind of peace between ourselves and a continually changing, variable environment. The path to this peace of mind is in the recognition that there can be a kind of harmony in processes and movement, just as there is in music: lyrics, melodies, themes. Like an opera, nature is made up of many events taking place simultaneously. In a theater drama without singing, only one speaker can be heard at a time, except for short periods that represent conflict, strife, or disorder, such as the mimicking of a war or many people speaking at a party. But in an opera or choral work, numerous voices can sing and be heard, understood, and appreciated. This can be explained physically in terms of the bandwidth of the simultaneous tones of multiple voices and through the mathematics of information theory and Fourier transforms. Yet the effect is not in these explanations but in the phenomenon itself. So it is with nature.

Among the failures in the human relationship with nature in modern times is a confusion about experts and expertise. Thoreau sought to understand and to connect with nature in every way he could. Not being able to experience everything himself, he relied at times on experts, but he did so carefully and in a limited way. The experts he sought were either the great thinkers and writers of Western history or contemporaries whose direct experience gave them intimate knowledge and understanding of the subject in question. Though he would dismiss those who spoke without experience, he could admire Alexander the Great carrying the Iliad; consult the Wellfleet Oysterman and the Old Woman on Cape Cod; admire Joe Polis, his third and best Indian guide; and seek to learn from loggers and fishermen, including those who fished for small whales. After listening to these experts, he would test for himself what they had told him. Expertise served as a starting point, not an end point, in his pursuit of nature-knowledge.

During a disaster, we may listen to experiential experts and even ask them questions inappropriate to their knowledge and understanding. In general, however, we tend to dismiss their value. An example can be found in the history and people of an area in New Jersey called the Pine Barrens. It is a large area and a curious one, an expanse of sandy soils and woodlands of small trees—pitch pines, scarlet oaks, white oaks—of blueberries and their relatives, of bracken ferns. The landscape is in some ways monotonous, but it has an appeal, especially along the streams that drain through it. Here, early European settlers found bog iron, commercially valuable rock- and boulder-sized pieces of iron ore produced by geological processes that take place in sandy soils near the sea. These settlers, who became known as "Pineys," have remained a relatively isolated group for generations. They are little known and much forgotten. In recent decades, environmentalists have become interested in conserving nature in the Pine Barrens, but at first they ignored the Pineys as irrelevant, unimportant, and having nothing to contribute to the solutions. Then, in the early 1990s, the staff of the American Folklife Center of the Library of Congress began to work with the Pineys. With the cooperation of these local people, they developed a traveling exhibition about their culture and history. About this, the Pineys were experiential experts. The project helped urban environmentalists recognize the value of the Pineys' culture and has led to solutions to environmental problems in the Pine Barrens that are beneficial to both nature and the local people. The urban environmentalists recognized that the local culture was worth preserving along with the ecological character of the Pine Barrens. The Pine Barrens did not become either an unpeopled nature preserve or an overdeveloped landscape stripped of ecological features and its indigenous culture.

Thoreau struggled throughout his life with the value of science as a path to making contact with nature. On one hand, he participated in scientific research and argued for the importance and utility of quantitative scientific measurements. On the other, he was never sure whether these activities aided his quest for a spiritual connection with nature. He vacillated about this. But he never accepted a lazy, unexamined approach to nature. The importance of measurement and of science to him is illustrated by his measurements of the depth of Walden Pond and the generalizations that he attempted from these.

A willingness to appreciate both civilization and nature leads us to a different perspective on cities. City environments should not be dismissed as bad places that we can only attempt to make less awful. Urban life will succeed when we recognize the positive potential of urban environmental settings. Wilderness, in turn, will benefit as cities are improved as human habitat. We need to ask ourselves what it is that maintains civilization. What gives civilization any permanence? If it is not our buildings and our technology, then what is it? It is the power of ideas, our heritage of knowledge.

The primary thrust of this book has been to suggest an alternative to today's conflicts between civilization and nature. This alternative lies *within* the rationalist tradition of recent Western civilization. It celebrates the products of civilization as well as the splendor of life on Earth. I have used Thoreau as a metaphor for a perspective and an approach. The key is to love both civilization and nature, to find fascination in nature and in the products of civilization, including its technology. If one follows the path traced by the history of Western civilization, with its combination of Greek-based rationality and the Judeo-Christian spiritual search, one comes to the approach I have suggested. It is the approach that Thoreau took in his daily walks around Concord—to the swamp on the edge of town—for renewal, inspiration, and fun.

Nature is no man's garden in the sense that it is not completely the product of human activities, yet what we appreciate as nature has been much influenced by our species. When Thoreau wrote that phrase at the top of Mount Katahdin he found nature "something savage and awful, though beautiful," a nature "made out of Chaos and Old night" that was "no man's garden," he was recognizing the integral place of people within nature, the idea that nature and human beings are one system, not separate, not separable. The search for everybody's garden leads us to places where people walk and with their footsteps leave different impressions in different places. In that swamp on the edge of town and in the more open spaces, like the forests of Maine, we can find the garden that will sustain nature *and* ourselves.

Notes

Epigraph

"Nature was here something savage and awful . . ." Henry David Thoreau, *The Maine Woods,* 2nd ed., edited by Joseph J. Moldenhauer (Princeton, N.J.: Princeton University Press, 1973), 70.

Introduction

Ron Arnold's Ecology Wars Ron Arnold, *Ecology Wars: Environmentalism as if People Mattered* (Bellevue, Wash.: Free Enterprise Press, 1993).

Andrew Rowell's Green Backlash Andrew Rowell, *Green Backlash: Global Subversion of the Environmental Movement* (London: Routledge, 1996).

"members of the general public concerned about the fate of the biosphere . . ." Thomas Dietz, professor of sociology and environmental science and public policy, George Mason University, personal communication based on his research and publications.

"We can be ethical only in relation to something we can see . . ." Aldo Leopold, *A Sand County Almanac: With Essays on Conservation from Round River* (New York: Oxford University Press, 1949; reprint, New York: Ballantine Books, 1970), 251.

"conservation is a state of harmony between men and land." Ibid., 243.

Chapter 1. Climbing Mount Katahdin

"Who are we? Where are we?" Henry David Thoreau, *The Maine Woods,* 2nd ed., edited by Joseph J. Moldenhauer (Princeton, N.J.: Princeton University Press, 1973), 59.

"I looked down the dizzy abyss . . ." William C. Larrabee, "The Backwoods Expedition," in *Rosabower: Essays and Miscellanies* (Cincinnati: Thompson, 1855), 253–272. Quoted in John W. Hakola, *Legacy of a Lifetime: The Story of Baxter State Park* (Woolwich, Maine: TBW Books, 1981), 13.

He hoped to achieve a spiritual connection with benign nature . . . The nature he sought to confront is what we refer to today as biological nature, as distinct from the cosmos.

Direct, personal experience is the most obvious way to learn about nature and one's connection to it, and it was Thoreau's primary approach to knowledge . . . The word *nature* is used to mean a variety of ideas and things, some much related to arguments about the existence of God. In Western civilization, there are three classic arguments for the existence of God: the perfection of the cosmos, the perfection of the anatomy of the human body, and the perfection of biological nature. Each of these—cosmos, human anatomy, and ecological systems—has been referred to as nature. In this book, for simplicity, I reserve *nature* for ecological systems and use specific terms for cosmos and anatomy. In this book, *nature* therefore includes references to what scientists call ecosystems, landscapes, the biosphere, populations, and species of wild organisms.

Mount Katahdin is the northern end of the Appalachian National Scenic Trail . . . Information about the Appalachian Trail and the height of Mount Katahdin is from the *Encyclopaedia Britannica CD '97*.

"The wood was chiefly yellow birch . . ." Thoreau, *Maine Woods,* 59.

"in a deep and narrow ravine . . ." Ibid., 60. Given that many people climb Mount Katahdin, some might like to know the route that Thoreau took in relation to modern trails. Thoreau approached the mountain from the south, roughly paralleling the present-day Abol Trail and about two miles to the east of it. He began by following Abol Stream; then he turned away from the stream and made his way up the mountain about one mile from what is today the Abol Bridge Campground, climbed just to the west (left) of what is now called Abol Mountain, and continued his ascent just to the west (left) of what is now called Rum Mountain. This brought him between South and Baxter Peaks. He reached the tableland between these peaks to the east of the modern Appalachian Trail. Information about his route is from J. Parker Huber, *The Wildest Country: A Guide to Thoreau's Maine* (Boston: Appalachian Mountain Club, 1981), 141–142.

"their tops flat and spreading . . ." (and other quotes from *The Maine Woods*) Thoreau, *Maine Woods,* 60–61.

"When I talk about the environmental movement . . ." Andrew Rowell, *Green Backlash: Global Subversion of the Environmental Movement* (London: Routledge, 1996), xi.

"the mainstream environmental movement . . ." Ibid., xii.

. . . responses of participants in a public opinion poll . . . The focus group was held by the Charlton Research Company of Walnut Creek, California. In December 1997, the Congressional Institute "commissioned this company to conduct a national research study to explore public judgment and values toward environmental issues." Previously, the company had conducted a comprehensive environ-

mental survey (1996) and a number of other environmental surveys. The methodology of the 1997 study included a telephone survey, lasting approximately thirty-two minutes, of 800 adults nationwide. According to the company's report, this sample size "was proportionate to the country's demographics including geography, gender, voter registration and ethnicity" and "yields a +/– 3.5% margin of error." The methodology also included a series of focus groups held in Salt Lake City, Utah; Fairfax, Virginia; and Farmington, Connecticut. Participants in the focus groups were eighteen years of age or older and met the Charlton Research Company's criteria for "informed Americans." Charles Rund, "Environmental Trends: Executive Summary" (Walnut Creek, Calif.: Charlton Research Company, winter 1998). The company's mailing address is 1460 Maria Lane, Suite 410, Walnut Creek, CA 94596.

Chapter 2. Crossing Umbazooksus Swamp

"I suspect that . . ." Henry David Thoreau, *The Maine Woods,* 2nd ed., edited by Joseph J. Moldenhauer (Princeton, N.J.: Princeton University Press, 1973), 13.

"With exaltation at the beauty of nature . . ." Clarence J. Glacken, *Traces on the Rhodian Shore: Nature and Culture in Western Thought from Ancient Times to the End of the Eighteenth Century,* reprint ed. (Berkeley: University of California Press, 1990), 707–708.

. . . his third and last trip . . . His two previous trips were in August–September 1846, when he climbed Mount Katahdin, and September 1843.

"In wildness is the preservation of the world . . ." Henry David Thoreau, *The Natural History Essays,* edited by Robert Sattelmeyer (Salt Lake City: Peregrine Smith Books, 1980), 112.

"all the baggage he had" Thoreau, *Maine Woods,* 159.

. . . baggage weighed 166 pounds. Ibid., 164.

"the wildest country" J. Parker Huber, *The Wildest Country: A Guide to Thoreau's Maine* (Boston: Appalachian Mountain Club, 1981).

"I had not much faith . . ." Thoreau, *Maine Woods,* 213.

. . . an idea that can be traced back to the Greeks and Romans and beyond. See Daniel B. Botkin, *Discordant Harmonies: A New Ecology for the Twenty-First Century* (New York: Oxford University Press, 1990).

"up to his neck and swimming for his life in his property, though it was still winter." Thoreau, "Walking," in *Natural History Essays,* 118.

Chapter 3. Enjoying the Swamp on the Edge of Town

"In Princeton College . . ." Larry McMurtry, *Streets of Laredo* (New York: Pocket Books, 1993), 261–262.

"Yes, though you may think me perverse . . ." Henry David Thoreau, "Walking," in *The Natural History Essays,* edited by Robert Sattelmeyer (Salt Lake City: Peregrine Smith Books, 1980), 114–115.

"It was a relief . . ." Henry David Thoreau, *The Maine Woods,* 2nd ed., edited by Joseph J. Moldenhauer (Princeton, N.J.: Princeton University Press, 1973), 155.

"It seemed to me that there could be no comparison . . ." Ibid., 155.

"The wilderness is simple . . ." Ibid.

"Our woods are sylvan . . ." Ibid.

"Perhaps our own woods and fields . . ." To this, Thoreau added, "They are the natural consequence of what art and refinement we as a people have,—the common which each village possesses, its true paradise, in comparison with which all elaborately and willfully wealth-constructed parks and gardens are paltry imitations." Ibid., 155–156.

"A town is saved . . ." Thoreau, "Walking," 117.

"The civilized nations . . ." Ibid.

Charles Goodyear developed the vulcanization process . . . Compton's Interactive Encyclopedia. Copyright © 1993, 1994 Compton's New Media, Inc.

. . . built a boat "in form like a fisherman's dory . . ." Henry David Thoreau, *Walden,* edited by J. Lyndon Shanley (Princeton, N.J.: Princeton University Press, 1971), 15.

Thoreau had graduated from Harvard University only two years earlier . . . Linck C. Johnson, historical introduction to *A Week on the Concord and Merrimack Rivers,* by Henry David Thoreau, edited by Carl F. Hovde, William L. Howarth, and Elizabeth H. Witherell (Princeton, N.J.: Princeton University Press, 1980), 433.

. . . a tribute and memorial to his brother . . . Ibid., 434.

"Gradually the village murmur subsided . . ." Thoreau, *Week on the Concord,* 19–20.

"Nature seemed to have adorned herself . . ." Ibid., 21.

"As naturally as the oak bears an acorn . . ." Ibid., 91.

"To read well . . ." and subsequent quotations from Thoreau in this chapter Thoreau, *Walden,* 100ff.

Chapter 4. Nurse Trees and Nature

"Others do not see what I see. . . ." *Los Angeles Times,* 7 March 1993, A21.

"The jay is one of the most useful agents . . ." Quoted in Henry David Thoreau, "The Succession of Forest Trees," in *The Natural History Essays,* edited by Robert Sattelmeyer (Salt Lake City: Peregrine Smith Books, 1980), 87.

Thoreau was stimulated by Darwin's book . . . Bradley Dean, personal communication.

He made direct observations about the ways in which seeds spread. Thoreau had a considerable background and a lifelong interest in the detailed study of nature and collection of specimens. About 1845, when he was starting his sojourn at Walden Pond, he began to study vegetation. He obtained a library of basic botanical guides. He learned plant taxonomy—the relatively new system for naming and describing plants developed by Carolus Linnaeus. He kept up with the work of his contemporaries who studied vegetation, including Asa Gray and Jean Louis Rodolphe Agassiz. Thoreau also studied fish in the rivers, streams, and lakes of Concord. He read books about insects and became acquainted with Thaddeus Harris, an entomologist at Harvard University. According to Robert Sattelmeyer, Thoreau "had been interested in ornithology since boyhood (a family album of bird sightings survives, dating from the 1830s and containing entries by Henry, his brother John, and his sister Sophia), and he compiled a large collection of birds' nests and eggs. In 1850 he was elected a corresponding member of the Boston Society of Natural History, to which he contributed specimens and various written accounts over the years, and whose library and collections he used regularly in pursuing his studies. (His own extensive collections of Indian artifacts, birds' nests and eggs, and pressed plants went to the Society after his death.) In his work as a surveyor, he made a more intimate acquaintance with the farms, swamps, and woodlots of Concord. Gradually his townsmen, who had generally looked askance at his activities, began to come around for help in identifying plants and animals, and to bring him new items for his collections." Sattelmeyer, introduction to *Natural History Essays,* xix–xx.

He measured the distance over which nesting birds and squirrels carried fruits and seeds . . . Henry David Thoreau, *Faith in a Seed: The Dispersion of Seeds and Other Late Natural History Writings,* edited by Bradley P. Dean (Washington, D.C.: Island Press, Shearwater Books, 1993), 28–29.

. . . transported by wind and water. Ibid., xv.

"the first Anglo-American field ecologist to be influenced by Darwin's theory." Ibid., xiv. Dean also stated that Thoreau's "observations on dispersal distances of fleshy fruits carried by animals remain important a century and a half later for their precision and anticipation of plant-animal mutualism."

"At first the young pines lined each side of the path like a palisade." Ibid., xiv.

. . . the succession of forest trees . . . Ibid., 33.

To my knowledge, the first use of the term *succession* to refer to the development of a forest or any ecosystem was made in 1784 by Thomas Pownall. He wrote: "The individual trees of those woods grow up, have their youth, their old age, and a period to their life, and die as we men do. You will see many a Sapling growing up, many an old Tree tottering to its Fall, and many fallen and rotting away, while they are succeeded by others of their kind, just as the Race of Man is. By this Succession of Vegetation the Wilderness is kept cloathed with Woods just as the human Species keeps the Earth peopled by its continuing Succession of Generations." Thomas Pownall, *A Topographical Description of the Dominions of the United States of America,* edited by Lois Mulkearn (1784; reprint, Pittsburgh: University of Pittsburgh Press, 1949). Thoreau, however, appears the first to apply the term within a more modern scientific context.

"The shade of a dense pine wood . . ." Thoreau, "The Succession of Forest Trees" (originally presented as an address to the Middlesex Agricultural Society in Concord, September 1860), in *Natural History Essays,* 78.

. . . benefit the oaks. See especially Thoreau, *Faith in a Seed.* See also Thoreau, "Succession of Forest Trees," 81. This essay is also the source of his statement that he has "faith in a seed." Dean's edited book *Faith in a Seed* includes previously unpublished material by Thoreau as well as this essay.

. . . modern environmental issues attract a wide range of opinions . . . As I touched on in chapter 1, current public debates about the environment can be separated into two categories, known informally as "brown" and "green" issues. Brown issues involve pollution of air, water, and soil. It would be hard to find someone who would argue for dirty air, water, or soil as a general good. Debates center not on goals but on how clean is "clean enough" and what amount of pollution would have no negative environmental effect on people. These debates continue to be quite sharp and are usually phrased in terms of how much pollution a society should accept in order to maintain global economic growth.

The green issues are biological: the question of what is wilderness and whether we need to, want to, or should save it; how to manage forests and fisheries; whether and how to conserve biological diversity and endangered species; why we need nature preserves and what their size, shape, and condition should be to enable them to serve their purposes; and what kinds of habitats, such as swamps and other wetlands, are especially in need of protection. In short, people generally agree that we should "clean up" where we have "messed up" the environment; what they disagree about is the value of the natural environment.

"Many contend that living beings can be ranked according to their relative intrinsic value." Arne Naess, *Ecology, Community, and Lifestyle: Outline of an Ecosophy,* translated by D. Rothenberg (New York: Cambridge University Press, 1989), 169. At the time he wrote this book, Naess was at the Council of Environmental Studies, University of Oslo. He is considered to be a primary philosopher of the deep ecology movement.

"The right of all the forms [of life] . . ." Ibid., 166. Naess does, however, admit to the need to eat. He states: "It is against my intuition of unity to say 'I can kill you because I am more valuable' but not against the intuition to say 'I will kill you because I am hungry.' In the latter case, there would be an implicit regret: 'Sorry, I am now going to kill you because I am hungry.' In short I find obviously right, but often difficult to justify, different sorts of behaviour with different sorts of living

beings. But this does not imply that we classify some as intrinsically more valuable than others" (168–169).

"No single species of living being has more of this particular right to live and unfold than any other species." Naess also admits that human beings have the unique ability to "consciously perceive the urge other living beings have for self-realization, and that we must therefore assume *a kind of responsibility for our conduct towards others*" (170). He refers to a "principle of biospheric egalitarianism defined in terms of equal right." He states that this has been misunderstood to mean that "human needs should never have priority over non-human needs. But this is never intended. In practice, we have for instance greater obligation to that which is nearer to us. This implies duties which sometimes involve killing or injuring non-humans" (170). He states that "the ecosophical outlook is developed through an identification so deep that one's *own self* is no longer adequately delimited by the personal ego or the organism. One experiences oneself to be a genuine part of all life. Each living being is understood as a goal in itself, *in principle* on an equal footing with one's own ego. It also entails a transition from I–it attitudes to I–thou attitudes—to use Buber's terminology" (174).

Interestingly, Naess does share with Thoreau the value of direct experience with nature through work and professions: "Wildlife and forest management, and other professions in intimate contact with nature, change people's attitudes. It is only through work, play, and understanding that a deep and enduring identification can develop, an identification deep enough to color the overall life conditions and ideology of a society" (176).

"the human drive for self-realization . . ." Ibid., 177.

"to relate all value . . ." Ibid., 178.

"The need for outdoor life . . ." Ibid., 182.

a rigid interpretation of ecological succession . . . I discuss the history of the idea of the balance of nature and its implications for modern ecology and environmental issues in Daniel B. Botkin, *Discordant Harmonies: A New Ecology for the Twenty-First Century* (New York: Oxford University Press, 1990).

Sorbonne professor of philosophy Luc Ferry . . . Luc Ferry, *The New Ecological Order*, translated by Carol Volk (Chicago: University of Chicago Press, 1995), 66.

the biosphere . . . Scientists today use the term *biosphere* to mean disparate things. One definition is the total amount of organic matter on the earth, in contrast with *hydrosphere*, all the earth's water; *lithosphere*, the bedrock of the continents and the tectonic plates; and *atmosphere*. Another definition is the global habitat of life—that is, the sum of the places where life is found, or the region that extends from the top of the troposphere to the depths of the oceans. Another definition, and the one I use in this book, is the global life-containing and life-supporting system—the meaning attached to the word by its originator, Vernasky. Most ecologists, however, use *biosphere* in the more limited meaning of the total amount of organic matter on the earth. In that case, they refer to the system that contains and supports life as the "earth system," which is confusing because it could mean many different things, including the entire planet with its core or the Earth–moon system.

. . . what Ferry calls a "strange hierarchy." Ferry, *New Ecological Order,* 66.

The simple notion . . . Ibid.

"the arrogance of stewardship . . ." Naess, *Ecology, Community, and Lifestyle,* 187.

Believers . . . dream of a "global government . . ." Ferry, *New Ecological Order,* 74. The quote is from Jean Fréchaut, in Prenoier and Le Seigner, *Génération verte* (Paris: Presses de al renaissance, 1992).

William Aiken, who asks whether . . . William Aiken, in *Earthbound: New Introductory Essays in Environmental Ethics,* edited by Tom Regan (New York: Random House, 1984), 269. Quoted in Ferry, *New Ecological Order,* 75.

"The deep ecology demand for the establishment of large territories . . ." Naess, *Ecology, Community, and Lifestyle,* 212. Naess continues: "Wild areas previously classified as 'voids' are now realized to be of vital importance and intrinsic value. This is an example of the kind of consciousness change that strengthens deep ecology. It must continue" (212). Once again, Naess misunderstands aspects of ecology and oversimplifies, as I discuss in later chapters. There is evidence, for example, that human-induced fires have helped maintain the Serengeti Plain and its great abundance of large mammals, whereas there is only weak evidence that large mammals cannot coexist with certain kinds of human presence and human activities. Naess ignores the potential for our technology to assist in the conservation of endangered species, as is being done, for example, in the transport of animals among preserves to maintain genetic diversity. Here, I argue that Naess demonstrates that a little knowledge is a dangerous thing. He seems to desire so strongly to portray technological civilization as an evil that he misreads the findings of ecology and personifies and anthropomorphizes what should be taken as scientific observations.

pollution "is an inevitable consequence . . ." Ron Arnold, *Ecology Wars: Environmentalism as if People Mattered* (Bellevue, Wash.: Free Enterprise Press, 1993), 116.

"Lovelock's clear-sighted vision . . ." Ibid., 116.

"we can love the Earth . . ." Ibid., 113, 115–116. Interestingly, in several instances, Arnold quotes Thoreau's famous statement "In wildness is the preservation of the world" as one of the earliest foundations for environmentalism. As I will show, Arnold, along with many of those he calls environmentalists, misunderstands Thoreau's meaning.

He is not the only one to misinterpret this aphorism by Thoreau. It is one of the most quoted of Thoreau's assertions. But Thoreau meant wildness as a state of mind, not a physical place, so the statement is erroneously used to justify the conservation or preservation of wilderness. Even so, this misinterpretation of Thoreau is legend. A recent example can be seen in a special issue of *Time* magazine, *Beyond 2000,* in an article titled "Will There Be Any Wilderness?" The author, Jon Krakauer, wrote: "In 1851, when Henry David Thoreau declared 'In wildness is the preservation of the world,' he could not have foreseen that wilderness, as an idea, would one day be used to sell everything from SUVs to soda pop."

Krakauer, like Arnold and so many others, confused Thoreau's wildness, a state of mind, with the physical entity of wilderness—a place, such as a mountaintop that Krakauer visited in Antarctica, which few had seen before and which lacked any evidence of the presence of human beings. He described his trip with two companions to Troll Castle, a piece of granite that rises "a vertical mile" above the ice cap: "My back hurt, and I had lost all feeling in my toes." It is unlikely that Thoreau would have appreciated wildness or wilderness in such a place. Jon Krakauer, "Will There Be Any Wilderness?" *Time,* 8 November 1999, 119–120.

Another person who presents a variation on this anti-environmental perspective is A. M. Gottlieb, in his introduction (pp. 7–8 especially) to Arnold's *Ecology Wars.* "These environmentalists," he wrote, believe that "everything we do leads to catastrophe, so everything ought to be stopped."

two-thirds of respondents considered themselves strong to moderate environmentalists . . . Charles Rund, "Environmental Trends: Executive Summary" (Walnut Creek, Calif.: Charlton Research Company, winter 1998), 2. This was a nationwide survey conducted by the Charlton Research Company of Walnut Creek, California. The sample size was 600 or more adults; the estimated margin of error was 4 percent.

Chapter 5. Racing in the Wilderness

"Thoreau did not think of the wild . . ." Robert Sattelmeyer, introduction to *The Natural History Essays,* by Henry David Thoreau, edited by Robert Sattelmeyer (Salt Lake City: Peregrine Smith Books, 1980), xxii.

"Barring love and war . . ." Aldo Leopold, *A Sand County Almanac: With Essays on Conservation from Round River* (New York: Oxford University Press, 1949; reprint, New York: Ballantine Books, 1970), 280.

Thoreau had "spent at least half the time in walking . . ." Quoted by J. Parker Huber in *The Wildest Country: A Guide to Thoreau's Maine* (Boston: Appalachian Mountain Club, 1981), 100.

"about a mile above Hunt's [house] . . ." Henry David Thoreau, *The Maine Woods,* 2nd ed., edited by Joseph J. Moldenhauer (Princeton, N.J.: Princeton University Press, 1973), 283.

"stopped early and dined . . ." Ibid., 284.

unable to obtain supplies . . . Huber, *Wildest Country,* 110. He notes that Hunt's inn had become a starting point for those who climbed Mount Katahdin; that mountain had become somewhat of a tourist attraction during the eleven years since

Thoreau climbed it. Among the visitors to the inn, Huber notes, was artist Frederic Edwin Church, who came there in 1855. By this time, Huber notes, a minister, Marcus R. Keep, who had first ascended Katahdin in the same year Thoreau did, was leading tourists up the mountain from Hunt's inn (107).

"When we had carried over one load . . ." Thoreau, *Maine Woods,* 285.

"I bore the sign of the kettle . . ." Ibid., 285–286.

"It is evident that he had a lot of fun in Maine." Huber, *Wildest Country,* 8.

"Staked to Fair Haven . . ." Henry David Thoreau, *Journal,* edited by Bradford Torrey, vol. 6 (Boston: Houghton Mifflin; Cambridge, Mass.: Riverside Press, 1906), 25.

This is not the image of Thoreau . . . Although Thoreau could obtain spiritual and creative inspiration from nature near Concord, and although he did not want to live in the wilderness of the Maine woods, far from civilization, he enjoyed going there. So do many people today. Recreation in both these senses is a perfectly good justification for the conservation of nature. Often, it brings economic benefit, as with ecotourism and other outdoor activities done in ways that are consistent with conservation practices, as skiing or golfing might be. The economic benefits of a sustainable natural area make up a large topic, beyond the scope of this book. One could argue that a large wilderness is desirable for recreation because hiking or canoeing through the vastness is itself aesthetically pleasing in the recreational sense—for physical pleasure, not for inspiration, and pleasing in a way that cannot be duplicated in smaller tracts.

"fully as bright as the fire . . ." Thoreau, *Maine Woods,* 179–180.

"It could hardly have thrilled me more . . ." Ibid., 179–181. I selected this among many of Thoreau's detailed experiences because I have seen the same phenomenon and been captivated by it. When I was in high school, I went on an overnight hiking trip to some nearby woods with a few friends. We dug a pit and built a fire in it, heated stones, and roasted potatoes. We listened to the sounds of the night and watched the stars through the trees. Then we set out our bedding and settled in. I awoke in the middle of the night to see phosphorescence around me. Unlike Thoreau, I had never heard of this phenomenon before, and it seemed quite magical to me. As Thoreau did, I got up and searched for the source of the glow. I found that it came from small twigs, which I gathered and put next to me so I could examine them in the morning. I was disappointed to see them in the morning as just a few more of the twigs on the ground, indistinguishable during the day from others that had not glowed. Today we know that this phosphorescence comes from a fungus, part of the group of organisms that decompose dead wood. This knowledge does not take away from the memory of the magic of that night.

. . . public service benefits of natural ecosystems. Some authors call the public service benefits of nature "ecological services" or "ecosystem services."

. . . the potential for finding new pharmaceuticals in the New World Tropics . . . Recently, a corporation called Shaman Enterprises, located in South San Francisco, sought to work with Indian tribes of tropical rain forests to find profitable pharmaceuticals from native plants and share the profits with the indigenous people who discovered the use.

The other person might be there because fishing . . . Public opinion polls conducted by the Charlton Research Company suggest that some people believe that contact with nature makes a person better; these respondents give this as the value of nature.

Without a clear idea of mission and purpose . . . Robert H. Nelson, "Rethinking Scientific Management," paper written for a conference of Resources for the Future. Nelson is a professor at the School of Public Affairs, University of Maryland.

. . . public opinion polls indicate that . . . This is seen in the results of public opinion polls conducted by Charles Rund of the Charlton Research Company (1460 Maria Lane, Suite 410, Walnut Creek, CA 94596).

It had been raining more or less for four or five days . . . Thoreau, *Maine Woods,* 173.

Known at that time as hornstone . . . Huber, *Wildest Country,* 28. The information here about the geology and Indian use of the rocks from Mount Kineo are from this reference, except for direct excepts from Thoreau, as noted.

Thoreau found a "small thin piece . . ." Thoreau, *Maine Woods,* 176.

Some Sioux Indians also present the argument . . . Personal communication, James Garrett, Cheyenne Sioux Indian and environmental scientist. Garrett, a friend of mine, is a graduate of the Environmental Studies Program at the University of California, Santa Barbara, and vice president of Cheyenne River Community College. James Garrett has proposed a plan to establish a bison and Indian migratory route between the U.S.–Canadian border and the Sioux Indian Reservations.

But the most constant and memorable sound of a summer's night . . . Henry David Thoreau, *A Week on the Concord and Merrimack Rivers,* edited by Carl F. Hovde, William L. Howarth, and Elizabeth H. Witherell (Princeton, N.J.: Princeton University Press, 1980), 40–41.

"evidence of nature's health." Ibid., 42.

Chapter 6. On Horseback Confronting the Great Desert

"One day I rode to a large salt-lake. . . ." Charles Darwin, *The Voyage of the Beagle,* in *The Harvard Classics,* vol. 29 (New York: Collier, 1909), 76.

"Let us faithfully record the impressions of the day . . ." Quoted in D. Lowenthal, *George Perkins Marsh: Versatile Vermonter* (New York: Columbia University Press, 1958), 137.

"by the red glare of the moon . . ." Edwin Bryant, *What I Saw in California* (Philadelphia: Appleton, 1848; reprint, with introduction by Thomas D. Clark, Lincoln: University of Nebraska Press, 1985), 170.

"straggling, stunted, and tempest-bowed cedars . . ." Bryant, *What I Saw* (1848), 172.

"dried crystals from the Great Salt Lake . . ." Thomas D. Clark, introduction to Bryant, *What I Saw* (1985), viii.

"About eleven o'clock . . ." Bryant, *What I Saw* (1848), 155.

"a scene so entirely new to us . . ." Ibid., 155–156.

"figures of a number of men and horses . . ." Ibid., 48.

"this phantom population . . ." Ibid., 117–118. Additional quotes from Bryant are from pp. 172–178.

"literary mission . . ." Clark, T. D., 1985, Introduction to the reprinting of Bryant's book, p. viii, University of Nebraska Press.

"a solitary wild rose . . ." Ibid., 48.

"completely surrounded by the forest as savage and impassible . . ." Henry David Thoreau, *Journal,* edited by John C. Broderick et al. (Princeton, N.J.: Princeton University Press, 1981–1997), vol. 2, *1842–1848,* 227, 311; quoted in R. D. Richardson, *Henry Thoreau: A Life of the Mind* (Berkeley: University of California Press, 1986), 179–182.

Chapter 7. Measuring the Pond

"Much has been learned since the end of the eighteenth century . . ." Clarence J. Glacken, *Traces on the Rhodian Shore: Nature and Culture in Western Thought from Ancient Times to the End of the Eighteenth Century,* reprint ed. (Berkeley: University of California Press, 1990), 707–708.

"Thoreau, like many other Romantics . . ." Robert Sattelmeyer, introduction to *The Natural History Essays,* by Henry David Thoreau, edited by Robert Sattelmeyer (Salt Lake City: Peregrine Smith Books, 1980), xii.

"There have been many stories . . ." and additional quotes about the depth of Walden Pond. Henry David Thoreau, *Walden,* edited by J. Lyndon Shanley (Princeton, N.J.: Princeton University Press, 1971), 285–289.

When scientists wish to test an inference . . . In this discussion of the scientific method, I am indebted to Dorothy B. Rosenthal, who wrote chap. 2 in Daniel B. Botkin and Edward A. Keller, *Environmental Science: Earth as a Living Planet,* 3rd ed. (New York: Wiley, 1999).

Thoreau speculated more broadly about possible generalizations . . . As Dorothy Rosenthal wrote: "The most typical hypotheses are those in the form of If . . . then statements. For example, 'If I apply more fertilizer, then the tomato plants will produce larger tomatoes.' The statement relates two conditions: amount of fertilizer applied and size of tomatoes. Each condition is called a variable, that is something that can vary. The size of tomatoes is called the dependent variable because it is assumed to depend on the amount of fertilizer, the independent variable. The independent variable is also sometimes called a manipulated variable, because the scientist deliberately changes or manipulates it. The dependent variable would then be referred to as a responding variable, one that responds to changes in the manipulated variable.

"In growing tomato plants, many variables may exist. Some, like the position of the North Star, can be assumed to be irrelevant; others, like the duration of daylight, are potentially relevant. In testing a hypothesis, a scientist tries to keep all relevant variables constant, except for the independent and dependent variables, an approach known as controlling variables. In a controlled experiment, the experiment is compared to a standard, or control, an exact duplicate of the experiment except for the one variable being tested (the independent variable). Any difference in outcome (dependent variable) between the experiment and the control can be attributed to the effect of the independent variable." Ibid., 21.

"Of course, a stream running through, or an island in the pond . . ." Thoreau, *Walden,* 289–290.

. . . must have altered his inner, spiritual relationship with them. Regarding this idea, "If we knew all the laws of nature," Thoreau wrote when he had completed this set of measurements, maps, and conjectures, "we should need only one fact, or the description of one actual phenomenon, to infer all the particular results at that point." *Walden,* 290–291.

Thoreau approached the study of ponds with the standard scientific method, which during his time was maturing in some fields of science . . . Henry David Thoreau, *Faith in a Seed: The Dispersion of Seeds and Other Late Natural History Writings,* edited by Bradley P. Dean (Washington, D.C.: Island Press, Shearwater Books, 1993). This volume includes Thoreau's participation in the question of whether there could be spontaneous generation of trees.

This scientific method is usually considered to have its roots in the end of the sixteenth and the beginning of the seventeenth centuries . . . According to D. Rosenthal, "We live in an age of science and high technology whose roots are ancient. Science had its beginnings in the ancient civilizations of Babylonia and Egypt, where observations of the environment were made primarily for practical reasons, as in determining the best location to plant crops, or religious reasons, as in using the positions of the planets and stars to predict the best time to plant. Ancient science differed from modern science in that it did not distinguish between science and technology or between science and religion.

These distinctions first appeared in classical Greek science. Because of their general interest in ideas, the Greeks developed a more theoretical approach to science, in which knowledge for its own sake became the primary goal. At the same time, their philosophical approach began to move science away from religion and toward philosophy. In fact, more than 100 years after the beginnings of modern science, well into the eighteenth century, science was still known as 'natural philosophy.'" In Botkin, D. B. and E. A. Keller, *Environmental Science: The Earth as a Living Planet,* John Wiley (New York), pp. 17-18.

Although not a practical scientist himself, Bacon recognized the importance of the scientific method . . . F. S. Taylor, *Science and Scientific Thought* (New York: Norton, 1949).

Geology was in an early developmental stage . . . Sir Charles Lyell, *Principles of Geology; Being an Attempt to Explain the Former Changes of the Earth's Surface by Reference to Causes Now in Operation* (1830–1833; reprint, New York: Johnson Reprint, 1969). Lyell's book was a great contribution. But in it, he dismissed the possibility that life could affect the earth's environment at a global scale, reasoning that living matter made up too small a fraction of the total material in the atmosphere, oceans, and upper layers of rock and soil. A few others in his time, however, held suspicions to the contrary. For example, by the time Thoreau was a teenager, the possibility that the burning of fossil fuels might lead to global warming had been proposed, obscurely.

. . . Darwin, Alfred Russell Wallace . . . Wallace was born in 1813. As a youth, he became interested in plants. In 1848, he traveled to the Amazon Basin, and in 1853 he published *A Narrative of Travels on the Amazon and Rio Negro.* In 1855, he wrote an essay titled "On the Law Which Has Regulated the Introduction of New Species." Like Darwin, he read *An Essay on the Principle of Population* by economist Thomas R. Malthus, and, like Darwin, he formulated from that essay the idea of survival of the fittest. Although he published the theory of evolution before Darwin, and he and Darwin presented the idea of biological evolution on the same day to the British Royal Society, Darwin has received most of the credit for it. This is in part because Darwin apparently studied and conceived of the theory earlier than did Wallace, but more so because Darwin worked out the theory in much greater detail. Nonetheless, Wallace was one of the major biologists of the nineteenth century.

The system for classifying plants and subsequently all forms of life . . . Linnaeus published his method for naming plants in *Genera plantarum* (1737).

. . . confusion was avoided, and biologists could build on one another's work. While writing this book, I helped direct small workshops on cancer research, which were attended by some of the leading scientists and clinicians in oncology. On a number of occasions, the discussions became confused because of disagreement among the experts about the definition of cancer. Some used the term *cancer* to mean a change in cell structure that could be detected under a microscope by a pathologist. Others assumed that the term applied only to a disease that spread throughout the body and was a cause of death. This confusion in terminology so interfered with the progress of discussions that some of the top people in the field circulated a defini-

tion among themselves, revising it until they arrived at something with which everyone was comfortable. This process of correspondence and discussion lasted for more than a year.

Thoreau read Darwin's account of his voyage on the Beagle . . . Charles Darwin, *Journal of Researches into the Geology and Natural History of the Various Countries Visited by H.M.S. Beagle* (London: Colburn, 1839).

"Thoreau's interest in science makes no sense apart from his Transcendentalist background." Thoreau, *Faith in a Seed,* 6. I am indebted to Dean for the discussion of Thoreau's study of the dispersal of seeds and its place in Thoreau's philosophy and epistemology.

Thoreau created what scientists call a model . . . C. M. Pease and J. J. Bull, "Is Science Logical?" *BioScience* 42 (1980): 293–298.

Although models are generally considered an integral part of the scientific method . . . The failure of ecologists to develop, test, and use theories and models has been discussed many times. I discussed the problem in two previous books, Daniel B. Botkin, *Discordant Harmonies: A New Ecology for the Twenty-First Century* (New York: Oxford University Press, 1990) (paperback edition published in 1993; Spanish translation published in 1993), and Daniel B. Botkin, *Forest Dynamics: An Ecological Model* (New York: Oxford University Press, 1993). There are other, classic discussions, such as K. E. F. Watt, "Use of Mathematics in Population Ecology," *Annual Review of Entomology* 7 (1962): 243–252. The specifics of the problems with models in ecological science are beyond the scope of this book.

Design an experiment to test the hypothesis. The development of a hypothesis requires identification of the dependent and independent variables. The dependent variable is the factor about which one wants to make a prediction. The independent variables are those factors that either (1) are correlated with the dependent variable to the extent that they can be used to predict the value of the dependent variable or (2) are actually the causes of the condition of the dependent variable.

"Our notions of law and harmony . . ." Thoreau, *Walden,* 290–291.

Thoreau's measurements of the depth of Walden Pond . . . A misunderstanding occurs when science is confused with technology. Science involves a search for understanding of the natural world, whereas technology involves control of the natural world for the benefit of humanity. Science often leads to technological developments, just as new technologies lead to scientific discoveries. The telescope, for example, began as a technological device, but when Galileo used it to study the heavens, it became a source of new scientific knowledge. That knowledge stimulated the technology of telescope making so that better telescopes were produced, and these in turn led to further advances in astronomy. People tend to confuse the products of science with science itself. Most of us do not come in direct contact with science in our daily lives; instead, we come in contact with the products of science—technological devices such as cars, toasters, and microwave ovens.

One of the most important misunderstandings about the scientific method is the relationship between research and theory. Although theories are usually presented as having grown out of research, in fact, theories also guide research. Scientists make their observations in the context of existing theories. At times, discrepancies between observations and accepted theories become so great that a scientific revolution occurs; old theories are discarded and replaced with new or significantly revised theories.

Their development demonstrates one way in which science is a process . . . T. S. Kuhn, *The Structure of Scientific Revolutions* (Chicago: University of Chicago Press, 1970). Dorothy Rosenthal wrote: "Evidence for models at the frontiers of science is not as strong as for those ideas accepted by the scientific community. With more research, some of these frontier models may move into the realm of accepted science and new ideas will take their place at the advancing frontier." J. S. Trefil, "A Consumer's Guide to Pseudoscience," *Saturday Review* 4 (1978): 16–21. However, research may not support other hypotheses at the frontier, and these will be discarded by scientists. Some people continue to believe in discarded scientific ideas, or pseudoscience. For example, although scientists long ago discarded the notion that movements of the heavenly bodies could affect people's personalities and personal fates, a very large proportion of Americans (estimates range from 40 to 50 percent) continue to believe in astrology. J. Miller, "The Scientifically Illiterate," *American Demographics* (June 1987): 26–31. However, much of what early astrologers learned about the movements of the stars and planets was so accurate that it became the basis of astronomy, the scientific study of the heavens. Some parts of astrology moved into the realm of accepted science; other parts, into pseudoscience. Rosenthal pointed out: "Accepted science may merge into frontier science, which in turn may merge into more far-out ideas, or fringe science. Really wild ideas may be considered beyond the fringe, or pseudoscientific. Although scientists have no trouble distinguishing between accepted science and pseudoscience, they do have trouble identifying ideas at the frontier that will become accepted and those that will be relegated to pseudoscience. That trouble arises because science is a process of continual investigation and an open system. This ambiguity at the frontiers of science leads many people to accept some frontier science before it has been completely verified and to confuse pseudoscience with frontier science." Botkin and Keller, *Environmental Science*, 25.

"I am pleased to learn that Thales was up and stirring by night not unfrequently . . ." Thoreau, *Natural History Essays*, 4–5.

"Science is always brave . . ." Ibid., 5.

"The regularity of the bottom and its conformity to the shores and the range of the neighboring hills were so perfect . . ." Thoreau, *Walden*, 289.

"The true man of science will know nature better by his finer organization . . ." Unreferenced quote from Thoreau in Sattelmeyer, introduction to *Natural History*

Essays, xv. Thoreau went on to say: "It is with science as with ethics,—we cannot know truth by contrivance and method; the Baconian is as false as any other, and with all the helps of machinery and the arts, the most scientific will still be the healthiest and friendliest man, and possess a more perfect Indian wisdom" (xv).

Meriwether Lewis touched on these aspects of natural beauty . . . The material on pesticides in the Missouri River is taken directly from Daniel B. Botkin, *Passage of Discovery: The American Rivers Guide to the Missouri River of Lewis and Clark* (New York: Penguin Putnam, Perigee Books, 1999).

By introducing large quantities of artificial compounds into our rivers, we are de facto conducting experiments on nature without the usual features of the scientific method . . . An important aspect of science, but one often overlooked in descriptions of the scientific method, is the need to define or describe variables in very exact terms that can be understood by all scientists. The least ambiguous way to define or describe variables is in terms of what one would have to do to duplicate them. Such definitions are called operational definitions because they use actions, or operations, rather than other terms. Thoreau's measurement of the depth of the pond is clearly an example of this definition of a variable.

. . . often it is because the standard scientific method either is not used or is used incompletely or incorrectly . . . Decision making with incomplete knowledge is a large topic and one that I do not pretend to address in this book. A common problem with environmental issues, it could itself be the subject of an entire book.

Chapter 8. The Poet and the Pencil

"The machine does not isolate man . . ." Antoine de Saint-Exupéry, *Wind, Sand, and Stars,* translated by Lewis Galantière (New York: Reynal and Hitchcock, 1939), 3.

"The speed with which Science marches from discovery to discovery . . ." Henry C. Kittredge, *Cape Cod: Its People and Their History* (Boston: Houghton Mifflin, 1930; reprint, Hyannis, Mass.: Parnassus Imprints, 1995), 3.

The pencil, first made of a thin rod of lead . . . This discussion of Thoreau and the pencil relies heavily on Henry Petroski's excellent book *The Pencil: A History of Design and Circumstance* (New York: Knopf, 1992). According to Petroski, the first reference to such a device is in a book about fossils titled *De Rurum Fossilium Lapidum et Gemmanrum Maxime, Figuris et Similitudinibus Liber,* written by Konrad Gesner in 1516.

Graphite found elsewhere . . . Petroski, *Pencil,* 64.

Varying the percentages of clay and graphite altered what we refer to today as the hardness of the lead . . . It is also possible, Petroski notes, that rather than developing this process independently, Thoreau came across a description of it in the *Encyclopaedia Britannica,* which he could have read in the public library.

. . . described this graphite-separating machine . . . The history of the Thoreau family's pencil business and Thoreau's participation in it are from Petroski, *Pencil.* The quotation by Emerson's son is from p. 114.

"over the following decade, during which time he was engaged on and off in the business . . ." Petroski, *Pencil,* 115. Other quotes from Petroski are from pages 116, 118, and 119.

. . . apparently inventing raisin bread . . . Petroski added: "But while he may not have won any ribbons for his cooking, in Thoreau's absence from the family pencil business, the Massachusetts Charitable Mechanic Association awarded a diploma. 'John Thoreau & Son for lead pencils exhibited by them at the exhibition and fair of 1847.'" Ibid., 121. He continued: "There is no mention of pencil making in *Walden,* but there is plenty of economics and sound thinking about business, qualities not alien to good engineering. Thoreau's famous accounting of the cost of the materials of his cabin ($28.12 1/2) and the profit he made from his 'farm' ($8.71 1/2) attests to his fondness and understanding of business." Ibid.

"scientific research is so specialized, and its hot fields so crowded, that it is highly unusual for an amateur to make a significant contribution." Nicholas Wade, "Amateur Shakes Up Ideas on the Recipe for Life," *New York Times,* 22 April 1997, C1. It is an irony of our time that creativity and excellence in thought are often determined by monetary criteria.

The first bicycle to use pedals, cranks, drive rods, and handlebars . . . *The Concise Columbia Encyclopedia* is licensed from Columbia University Press. Copyright © 1995 by Columbia University Press. All rights reserved.

The pneumatic tire . . . *The Concise Columbia Encyclopedia* is licensed from Columbia University Press. Copyright © 1995 by Columbia University Press. All rights reserved.

. . . the patent on James Watt's steam engine . . . *The Concise Columbia Encyclopedia* is licensed from Columbia University Press. Copyright © 1995 by Columbia University Press. All rights reserved.

. . . the motivation for ecosystem research . . . In the late 1950s and early 1960s, the Atomic Energy Commission began to fund three big projects to research the effects of radioactive elements on ecosystems, one at Brookhaven National Laboratory, one at Oak Ridge National Laboratory, and one in Puerto Rico. At Brookhaven, cesium 137, a source of gamma rays, was placed in the middle of an oak and pine forest on the laboratory's lands. At Oak Ridge, alpha- and beta-emitting elements were introduced into ecological food chains and their fate traced. In Puerto Rico, a source of gamma radiation was flown into a tropical rain forest for a short time so scientists could study the effects of acute radiation exposure on a forest. The funds applied to these studies dwarfed previous funding for the study of ecosystems.

According to William H. Cook . . . William H. Cook, *The Road to the 707: The Inside Story of Designing the 707* (Bellevue, Wash.: TYC, 1991), 3.

By the time of the development of the Boeing 707 . . . The information and history of the development of the jetliner presented here are from Cook, *Road to the 707,* 275. Cook notes that the gas turbine engine, originally patented in England in 1791, was first built in France in 1906. However, it was only 3 percent efficient and could not maintain the movement of its own parts. In 1907, the General Electric Company considered gas turbines as an emergency replacement for steam turbines (99), but they were again found too inefficient. Whittle and Ohain invented ways to make the gas turbine efficient.

One of the first French pilots, Saint-Exupéry . . . Antoine (-Marie-Roger) de Saint-Exupéry was born in 1900 and died in 1944, while flying a reconnaissance aircraft during World War II. He received his pilot's license in 1922 and began to fly airmail routes for France in 1926, first to North Africa and then in South America. Later, he worked as a newspaper reporter in France. In the United States, he is best known for his book *Le Petit Prince* (*The Little Prince,* 1943), a children's story with a message for adults, written while he was living in the United States at the beginning of World War II. He wrote of the beauty of the earth, the importance of human relationships, and the value of civilization.

Joseph Conrad . . . Conrad was born in what was then Poland but is now part of Ukraine, in 1857, when Thoreau was traveling to Cape Cod and the Maine woods. Born Józef Teodor Konrad Korzeniowski, he anglicized his name after moving to England and changing his career from ship's officer to novelist. His works are, of course, well known, and some of them are assigned as reading in college as frequently as is *Walden.* Conrad died in 1924.

Chapter 9. Breakfasting on Cape Cod

"The Cape is merely a temporary deposit . . ." Quoted in G. O'Brien, ed., *A Guide to Nature on Cape Cod and the Islands* (Hyannis, Mass.: Parnassus Imprints, 1990), 28.

He integrated all of these trips into a single narrative in his book Cape Cod. Thoreau used the journals he kept on his first visit to Cape Cod as the basis for lectures in 1849–1850; then he wrote a narrative of this visit and submitted it to *Putnam's Monthly,* in which it was published in 1855. This included the first four chapters of what would later become the book. The final book, with ten chapters, was published three years after Thoreau's death, in 1865. Joseph J. Moldenhauer, historical introduction to *Cape Cod* by Henry David Thoreau, edited by Joseph J. Moldenhauer (Princeton, N.J.: Princeton University Press, 1988), 249.

"uncommonly near the eastern coast." Thoreau, *Cape Cod,* 62.

"a grizzly-looking man appeared." Ibid., 63.

. . . his name was John Young Newcomb. Moldenhauer, historical introduction to Thoreau, *Cape Cod,* 250.

"the merriest old man . . ." Thoreau, *Cape Cod,* 71.

"repeatedly injured." Thoreau, *Cape Cod,* 77–78.

. . . he had seen cedar stumps . . . Thoreau, *Cape Cod,* 120–121.

"According to the light-house keeper . . ." This and subsequent quotes about the dynamics of the Cape in Thoreau, *Cape Cod,* 118–122.

. . . the Cape was not considered to be the beautiful recreational destination it is today. One expert on Thoreau's writing, Dana Brown of the University of Vermont, suggests that Thoreau's book *Cape Cod* is at one level an "anti-travel" travel book, a parody of the travel books that were just then becoming popular—and becoming possible because of vast improvements in transportation brought about by railroads, steamships, and better roads and bridges.

"Sand is the great enemy here . . ." Thoreau, *Cape Cod,* 173–179.

"Witness Stand: Who is an Expert? . . ." R. B. Schmitt, "Witness Stand: Who Is an Expert? In Some Courtrooms, the Answer Is 'Nobody,'" *Wall Street Journal,* 17 June 1997, 1.

Chapter 10. The Sound of a Woodchopper's Ax

"The individual trees of those woods grow up . . ." Thomas Pownall, *A Topographical Description of the Dominions of the United States of America,* edited by Lois Mulkearn (1784; reprint, Pittsburgh: University of Pittsburgh Press, 1949).

"What though the woods be cut down." Henry David Thoreau, "Huckleberries," in *The Natural History Essays,* edited by Robert Sattelmeyer (Salt Lake City: Peregrine Smith Books, 1980), 227.

. . . Joe Aitteon, an Indian guide the two had hired. Thoreau referred to Joe Aitteon as "a son of the Governor" who "had conducted two white men a-moose-hunting in the same direction the year before." Henry David Thoreau, *The Maine Woods,* 2nd ed., edited by Joseph J. Moldenhauer (Princeton, N.J.: Princeton University Press, 1973), 85. Later, he described him as "a good looking Indian, twenty-four years old, apparently of unmixed blood, short and stout, with a broad face and reddish complexion, and eyes, methinks, narrower and more turned up at the outer corners than ours. . . . He wore a red flannel shirt, woolen plants, and a black Kosuth hat, the ordinary dress of the lumberman, and to a considerable extent, of the Penobscot Indians." Ibid., 90.

But then, Thoreau "suddenly saw the light and heard the crackling of a fire . . ." Thoreau, *The Maine Woods,* Moldenhauer edition, pp. 100–101 The complete quote is: "I often wished since that I was with them. They search for timber over a given section, climbing hills and often high trees to look off,—explore the streams by which it is to be driven, and the like,—spend five or six weeks in the woods, they two alone, a hundred miles or more from any town,—roaming about, and sleeping on the ground where night overtakes them,—depending chiefly on the provisions they carry with them, though they do not decline what game they come across,—and then in the fall they return and make report to their employers, determining the number of teams that will be required the following winter. . . . It is a solitary and adventurous life, and comes nearest to that of the trapper of the West, perhaps. They work ever with a gun as well as an axe, let their beards grow, and live without neighbors, not on an open plain, but far within a wilderness."

"Fishermen, hunters, woodchoppers, and others, . . ." Henry David Thoreau, *Walden,* edited by J. Lyndon Shanley (Princeton, N.J.: Princeton University Press, 1971), 210.

"Very few men can speak of Nature with any truth." Henry David Thoreau, *A Week on the Concord and Merrimack Rivers,* edited by Carl F. Hovde, William L. Howarth, and Elizabeth H. Witherell (Princeton, N.J.: Princeton University Press, 1980), 108–109.

"soon confused by numerous logging-paths" This and subsequent quotes about the character of wilderness in Thoreau, *Maine Woods,* 212–213.

. . . a small steamboat called The Captain King *. . .* Thoreau, *Maine Woods,* 89–94.

. . . he came across a stack of hay that had been put there for a lumberman to use during the winter . . . Thoreau, *Maine Woods,* 106.

On Tuesday, July 28, 1857 . . . Thoreau, *Maine Woods,* 227–228.

"Humboldt has written an interesting chapter . . ." Thoreau, *Maine Woods,* 151.

. . . the earth would be "comparatively bare and smooth and dry." Thoreau, *Maine Woods,* 151–154.

. . . then "[left] the bears to watch the decaying dams. . . ." Thoreau, *Maine Woods,* 228.

. . . wildness was *a spiritual state . . .* For further discussion of the American idea of wilderness, see Roderick Nash, *Wilderness and the American Mind, 3rd ed.* (New Haven, Conn.: Yale University Press, 1982).

"The Anglo-American can indeed cut down, and grub up all this waving forest . . ." Thoreau, *Maine Woods,* 229.

"I caught a glimpse of a woodchuck . . ." Thoreau, *Walden,* 210.

Wilderness, the physical entity . . . For an important discussion of related ideas, see Nash, *Wilderness.*

Chapter 11. Finding Salmon on the Merrimack

"Science itself . . ." Joseph Campbell, *The Masks of God: Primitive Mythology* (New York: Viking Press, 1959; reprint: New York: Penguin Books, 1987), 468.

"It is the marriage of the soul with Nature . . ." Henry David Thoreau, *Journal,* 21 August 1851.

. . . river had banks "generally steep and high" Henry David Thoreau, *A Week on the Concord and Merrimack Rivers,* edited by Carl F. Hovde, William L. Howarth, and

Elizabeth H. Witherell (Princeton, N.J.: Princeton University Press, 1980), 86. A major reason why one might consider Thoreau's perspective on the utilization of forests, allowing a combination of civilization and nature, intensive use of natural resources, and an ability to experience spiritual uplifting within a humanly altered natural setting, inappropriate for our time is the great alternations we have imposed on nature worldwide since Thoreau's time. Perhaps, one might think, Thoreau was generally unaware of the loss of species or even of the possibility of the loss of species or major alterations of the environment. This did not, however, appear to be the case with the regional alteration of forests in New England. Of these Thoreau seemed well aware, as discussed later in this chapter.

. . . the river descended "over a succession of natural dams . . ." Ibid., 86–88.

Other explorations . . . Those interested in this history of the study of natural history in North America will profit from Susan Delano McKelvey's *Botanical Exploration of the Trans-Mississippi West, 1790–1850* (Corvallis: Oregon State University Press, 1991).

William Derham The material about Derham and Bernardin de Saint-Pierre that follows is from Daniel B. Botkin, *Discordant Harmonies: A New Ecology for the Twenty-First Century* (New York: Oxford University Press, 1990). Its relevance makes repetition of these ideas useful here.

. . . Physico-Theology: . . . William Derham, *Physico-Theology; or, A Demonstration of the Being and Attributes of God from His Works of Creation* (London: Strahan, 1798). The original edition included the statement that this work was "the substance of sixteen sermons, preached in St. Mary-le-Bow Church, London; at the Honourable Mr. Boyle's lectures, in the years 1711 and 1712."

He struggled with several issues . . . These issues are thoroughly discussed by Clarence J. Glacken in his excellent book *Traces on the Rhodian Shore: Nature and Culture in Western Thought from Ancient Times to the End of the Eighteenth Century* (Berkeley: University of California Press, 1967). Another classic and important book on this topic is A. O. Lovejoy, *The Great Chain of Being* (Cambridge, Mass.: Harvard University Press, 1942). Other interesting analyses can be found in F. N. Egerton, "Changing Concepts of the Balance of Nature," *Quarterly Review of Biology* 48 (1973): 322–350. The discussion that follows merely outlines the history of these ideas; readers interested in this history should refer to these references especially.

"By a curious harmony . . ." Glacken, *Traces,* 257–259.

Jacques-Henri Bernardin de Saint-Pierre Quoted in Egerton, "Changing Concepts," 338.

Thomas Robert Malthus Malthus was born in Dorking, England, in February 1766 and died near Bath, England, in December 1834. His famous essay went through a number of editions, each expanding and developing his ideas, but it became well known immediately after its first publication. His father was a friend of philosopher David Hume. Malthus excelled in his studies and became a professor of history and political economy at the East India Company's college at Haileybury, Hertfordshire. *Encyclopaedia Britannica CD '97.*

"Passion between the sexes . . ." Thomas R. Malthus, *An Essay on the Principle of Population, as It Affects the Future Improvement of Society, with Remarks on the Speculations of Mr. Godwin, M. Condorcet, and Other Writers* (London: Johnson, 1798; reprint, as *The First Essay on Population,* London: Royal Economic Society, 1927).

Between 1840 and 1920, with great precision and economic frugality, whalers hunted the mighty bowhead whale. J. R. Bockstoce and D. B. Botkin, *The Historical Status and Reduction of the Western Arctic Bowhead Whale (Balaena mysticetus) Population by the Pelagic Whaling Industry, 1848–1914,* final report to the National Marine Fisheries Service prepared by the Old Dartmouth Historical Society, New Bedford, Mass., 1980; J. R. Bockstoce and D. B. Botkin, "The History of the Reduction of the Bowhead Whale *(Balaena mysticetus)* Population by the Pelagic Whaling Industry, 1848–1914," in *International Whaling Commission Reports,* Special Issue 5 on historical whaling records, edited by M. F. Tillman and G. P. Donovan, 1983); D. A. Woodby and D. B. Botkin, "The Original Abundance of the Bowhead Whale," chap. 10 in *The Bowhead Whale,* Special Publication 2, edited by J. J. Burns, J. J. Montague, and C. J. Cowles (Lawrence, Kans.: Society for Marine Mammalogy, 1992).

John Jacob Astor *The American Heritage Dictionary of the English Language,* Third Edition copyright © 1992 by Houghton Mifflin Company. Electronic version licensed from INSO Corporation. All rights reserved.

"I have stood under a tree in the woods . . ." Thoreau, *Week on the Concord,* 299–300.

. . . Thoreau's approach to knowledge of nature remains a key to sustaining nature and civilization together. Structurally, nature is different. But the processes required to sustain natural systems are the same, and the paths to knowledge of and contact with nature involve the same processes that Thoreau found for himself. It is on these processes that we must focus if we are to sustain ourselves, our love of nature, and nature itself. Another way to put this is to say that Thoreau's approach will seem out of date if we focus on structure but not if we focus on process.

Chapter 12. Putting Forests on the Ballot

"Men and Nature must work hand in hand. . . ." Inscription on walkway at the Franklin Delano Roosevelt Memorial, Washington, D.C., from a message to the United States Congress on use of natural resources, 24 January 1935.

"Our minds anywhere . . ." Henry David Thoreau, *The Maine Woods,* 2nd ed., edited by Joseph J. Moldenhauer (Princeton, N.J.: Princeton University Press, 1973), 277.

In 1990, forests and other wooded lands . . . Information about the condition of forests worldwide is from Daniel B. Botkin and Edward A. Keller, *Environmental Science: Earth as a Living Planet,* 3rd ed. (New York: Wiley, 1999); R. A. Sedjo and D. B. Botkin, "Using Forest Plantations to Spare the Natural Forest," *Environment* 39, no. 10 (1997): 14–20.

An organization called Common Sense for Maine Forests . . . Common Sense for Maine Forests, "Strange Bedfellows," available on-line at http://www.mainecommonsense.com/bedfllows.html.

"the most expensive referendum in the state's history . . ." Associated Press, 6 November 1996.

Sayen saw this issue as a political and public relations battle . . . J. Sayen, *Earth Island Journal* (spring 1996).

. . . he was taken to the site of a clearcut. John M. Hagan III, "Clearcutting in Maine: Would Somebody Please Ask the Right Question?" *Maine Policy Review* (July 1996): 7–19. The quotations in the discussion that follows are from this article.

He decided to follow Haag's advice, reasoning, "Once I had documented that clearcuts were avian deserts . . ." Since Hagan's research project began, the land has changed hands several times. At the time of this writing (January 2000), it was owned by the Plum Creek Timber Company.

These observations led him to conclude . . . I discussed these events with Haag, who verified them and added much to them, emphasizing Hagan's positive and constructive efforts.

These observations led him to conclude . . . Hagan, "Clearcutting in Maine."

John Perlin John Perlin, *A Forest Journey: The Role of Wood in the Development of Civilization* (New York: Norton, 1989). Perlin makes the same point about the contradictory attitudes toward forests that have been dominant through Western history.

. . . in the Anglo-Saxon epic poem Beowulf *. . .* The word *wilderness* is from the Anglo-Saxon *wild doer,* "the place of wild creatures."

In a forested landscape, a variety of biological structural types . . . In the rush to save biological diversity, there is a tendency to conserve areas of local high diversity. However, the overall conservation of biological diversity requires conservation of some areas of low diversity as well because these are the habitats of a few but endangered or threatened or rare species.

Chapter 13. Baxter and His Park

"Man is born to die. . . ." Quoted in John W. Hakola, *Legacy of a Lifetime: The Story of Baxter State Park* (Woolwich, Maine: TBW Books, 1981), frontispiece. Baxter pur-

chased, with his own funds, the lands that include Mount Katahdin and gave them to the state of Maine.

"The kings of England formerly had their forests . . ." Henry David Thoreau, *The Maine Woods,* 2nd ed., edited by Joseph J. Moldenhauer (Princeton, N.J.: Princeton University Press, 1973), 156.

Otherwise, we might "grub them all up . . ." Ibid.

. . . national parks are an American idea. Alfred Runte presents the history of the national park in his *National Parks: The American Experience* (Lincoln: University of Nebraska Press, 1979).

The word parc *. . .* Ibid., 2.

. . . a single Public Reserved Lands system . . . R. H. Gardiner, "Public Reserved Lands of Maine: Planning Policy," report prepared for the Maine Department of Conservation, Bureau of Public Lands, 1985.

The present guiding principle of the Public Reserved Lands . . . ME. REV. STAT. ANN. tit. 12, sec. 556, quoted in Gardiner, "Public Reserved Lands," 3.

Baxter eventually purchased 200,000 acres . . . Hakola, *Legacy of a Lifetime,* 73. Information about Baxter State Park presented here is from this volume. The park is administered by the Baxter State Park Authority, Millinocket, Maine, (207) 723 5140. Subsequent quotes from Baxter about the park are from this source.

. . . the present director of the park . . . Buzz Caverly, telephone conversation with the author, 5 November 1999.

Chapter 14. Creating Wilderness

"Just before night we saw a musquash. . . ." Henry David Thoreau, *The Maine Woods,* 2nd ed., edited by Joseph J. Moldenhauer (Princeton, N.J.: Princeton University Press, 1973), 206–207.

A 1964 federal law, the Wilderness Act, defines wilderness . . . As defined in the Wilderness Act, Public Law 88-577 (16 U.S.C. 1131–1136), 88th Congress, 2nd session, 3 September 1964: "A wilderness, in contrast with those areas where man and his works dominate the landscape, is hereby recognized as an area where the earth and its community of life are untrammeled by man, where man himself is a visitor who does not remain. An area of wilderness is further defined to mean in this Act an area of undeveloped Federal land retaining its primeval character and influence, without permanent improvements or human habitation, which is protected and managed so as to preserve its natural conditions and which (1) generally appears to have been affected primarily by the forces of nature, with the imprint of man's work substantially unnoticeable; (2) has outstanding opportunities for solitude or a primitive and unconfined type of recreation; (3) has at least five thousand acres of land or is of sufficient size as to make practicable its preservation and use in an unimpaired condition; and (4) may also contain ecological, geological, or other features of scientific, educational, scenic, or historical value."

Thoreau rarely mentions Mount Katahdin elsewhere . . . Elizabeth H. Witherell, editor-in-chief of the Thoreau Edition, personal communication.

"I would not have every man nor every part of a man cultivated . . ." Henry David Thoreau, "Walking," in *The Natural History Essays,* edited by Robert Sattelmeyer (Salt Lake City: Peregrine Smith Books, 1980), 126.

"I seek the darkest wood . . ." Ibid., 116–117.

A society cannot maintain very large wilderness areas and simultaneously allow unlimited human population growth. Here, I use the word *allow* not in the restricted sense of government regulation but in its broadest interpretation, including social mores functioning at the level of individual and community cultural practices.

"Life consists with wildness. . . ." Thoreau, "Walking," 114.

"There is something in a strain of music . . ." Ibid., 121–122. In this same passage, Thoreau wrote, "Give me for my friends and neighbors wild men, not tame ones." But once again, this has to be interpreted in light of the way Thoreau saw wildness—as an internal quality, not an external quantity. A "wild man" might be a neighbor, a forester, a fisherman, a person in a cabin in the woods near town. Being "wild and free" did not, for Thoreau, require vast areas, even though if taken out of context, this statement might seem to argue for this.

"In literature it is only the wild that attracts us . . ." Ibid., 119.

Chapter 15. Conserving Wilderness

"I wish to speak a word for Nature . . ." Henry David Thoreau, "Walking," in *The Natural History Essays,* edited by Robert Sattelmeyer (Salt Lake City: Peregrine Smith Books, 1980), 93.

. . . the establishment of a Maine National Park . . . One rationale for establishment of the park given by RESTORE: The North Woods is that its members consider the forests of Maine to be mismanaged at present and place the blame primarily on large international corporations. The organization's newsletter states: "For decades the private landowners took reasonably good care of the forest, allowed open access to the land, and provided jobs to local people. In recent years, a handful of transnational corporations based in places such as Idaho, Canada, Japan, and South Africa have gained control of the region. Since 1970, they have clearcut an area larger than Delaware, sprayed millions of acres with pesticides, built 25,000 miles of logging roads, and subdivided tens of thousands of acres for lakeshore development. Not only has the forest suffered, but so have local people as these companies are eliminating hundreds of jobs through downsizing and mechanization. They are leaving a legacy of depleted forests, endangered wildlife, polluted rivers, and declining economies." Michael Kellett, "Why a Maine Woods National Park?" *North*

Woods Vision: The Newsletter of RESTORE: The North Woods (1996): 2. Kellett is the organization's executive director.

Walking at a rate of 2 miles per hour . . . Twenty miles per day is a good distance to travel in wilderness, as attested to by the experiences of Meriwether Lewis and William Clark. Eighteen miles was a good day's travel up the Missouri River for them. Experienced outdoorsmen, such as Dale Whitmore, a professional fishing and hunting guide who has guided people for years in Alaska and Arizona, have found twenty miles to represent a good day's cross-country travel. It has been my experience as well.

. . . a series of heavily wooded parallel ridges and valleys, in which one could follow the valleys or ridges. This is something like the design of a golf course, in which the holes are fit like pieces of a puzzle into several hundred acres. However, a designed wilderness is not what is sought and, in fact, would seem in modern parlance to be an oxymoron. But a golf course is typically 18,000–21,000 feet (3.4–3.9 miles) long and, if formed into a circle, would have a diameter of 1.08–1.24 miles and an inside area of 0.9–1.2 square miles, or 576–768 acres: 2.9–3.8 times as much area as is usually used. A similarly shaped wooded wilderness made up of parallel ridges and valleys would therefore appear to fit into an area one-third to one-fourth of the circular shape. Such a wilderness might then require approximately 1 million to 1.3 million acres, about the size of Boundary Waters. We can take this as a minimum area for a Leopold-style wilderness.

It would be convenient if scientific information gave us a clear idea . . . The proposal states that the park would include areas that the authors of the proposal consider to be of ecological value, including old-growth and second-growth forests, "hundreds of pristine lakes and ponds, and adequate habitat for the full range of native wildlife." The park would also protect cultural heritage "rich with Native American, colonial, and logging-era settlements, sites and travel routes."

. . . for which there is no adequate scientific justification. Formal statistical and scientific methods have recently been developed to determine what kinds of lands and what locations would be necessary to develop a representative set of natural areas. This is one of the topics for which an adequate scientific base is now possible.

This is inconsistent with the scientific method . . . Of course, in practice one estimates the minimum area as a lower bound to be avoided. One would try to achieve an area larger than the minimum, as a safeguard against unforeseen future events. The calculation of the minimum is important so that one knows where the lower bound lies, but it is not the desirable goal, which would always be greater than this minimum.

If predators such as wolves are necessary . . . I discuss the history of the idea that predators control the abundance of prey and the lack of evidence that this is true in Daniel B. Botkin, *Discordant Harmonies: A New Ecology for the Twenty-First Century* (New York: Oxford University Press, 1990).

Following their introduction into new habitats, . . . G. Caughley, "Eruption of Ungulate Populations, with Emphasis on Himalayan Thar in New Zealand," *Ecology* 51 (1970): 53–72, provides an analysis of case histories demonstrating that ungulates undergo population eruptions when introduced into new habitats, with or without the presence of predators.

The argument for this number is limited to genetic requirements . . . The primary genetic problem considered here is prevention of genetic drift—the fixation of a limited set of characteristics that might not be adapted over the long term to the environmental conditions of the available habitats. There is a need for additional research and an expansion beyond genetic concerns.

. . . the average home range for adult male wolves in Wisconsin . . . Adrina P. Wydeven, Ronald N. Scultz, and Richard P. Thiel, "Monitoring of a Recovering Gray Wolf Population in Wisconsin, 1879–1991," in *Ecology and Conservation of Gray Wolves in a Changing World,* edited by L. N. Carbyn, S. H. Fritts, and D. R. Seip (Edmonton: University of Alberta Press, 1995), 148. The average home range is reported in metric units as 179 square kilometers per adult male wolf year-round.

The area and population required to sustain wolves in Maine could be reduced further. If one's goal for a wilderness is the conservation of total species diversity, one useful approach is to determine the species area curve, a detailed discussion of which is beyond the scope of this book. Briefly, suppose in our visit to Big Reed Pond we made a list of the tree species we encountered, marking off plots as we went and naming the species we had not yet seen. In the first plot, all the species would be new to the list. In the second, we would find some repeats and some new species. The more plots we measured, the fewer new ones we would find. Gradually, the number of species new to our study would decline. We could plot the number of new species encountered versus the area sampled. This approaches a constant total number of species. Its asymptote is the estimate of the total number of species in an area and the total area required to contain all those species. But this method has its limitations. We started our walk through uplands, and we would get one species area curve for that type of forest. Then we descended into wetlands, and there a new species area curve could be established—suddenly species we had not seen would appear to us. The species area curve is sensitive to the size of the typical species (consider ants versus trees). Also, this method looks only at structural features, not processes.

It is also worth noting that considerable effort is being made in various lines of scientific investigation regarding the minimum size of viable, self-sustaining areas. A full discussion of these exceeds the scope of this book, but the main point is that our need to know remains ahead of our scientific understanding.

. . . a fascinating biography of the Wright Brothers . . . Harry Combs and Martin Caidin, *Kill Devil Hill: Discovering the Secret of the Wright Brothers* (Englewood, Colo.: TernStyle Press, 1979), 53–54. Combs, an experienced and early flyer and later head of Learjet Corporation, understood flying and the problems of flight intimately.

His perspective on the history of flight parallels my view of Thoreau's work and the relationship between humanity and nature, based on my experience in ecology and environmental sciences.

Jym St. Pierre gave further justifications for the park . . . Jym St. Pierre, "A Wildland under Siege," *North Woods Vision: The Newsletter of RESTORE: The North Woods* (1996): 3.

Jon Luoma describes cross-country skiing . . . Jon Luoma, "Recreation in the Maine Woods," *North Woods Vision: The Newsletter of RESTORE: The North Woods* (1996): 3

Chapter 16. Viewing the Ocean as Nature

"Like the sea itself, . . ." Quoted in G. O'Brien, ed., *A Guide to Nature on Cape Cod and the Islands* (Hyannis, Mass.: Parnassus Imprints, 1990), 74.

"It was something formidable and swift, . . ." Joseph Conrad, "Typhoon," in *A Conrad Argosy* (Garden City, N.Y.: Doubleday, Doran, 1942), 239.

"I have spent, in all, about three weeks on the Cape. . . ." Henry David Thoreau, *Cape Cod,* edited by Joseph J. Moldenhauer (Princeton, N.J.: Princeton University Press, 1988), 3.

"What is the sea to a fox?" Ibid., 146.

"a most advantageous point . . ." Ibid., 147.

Thoreau wrote much about the dead on the shore, . . . The quotations about the debris from the ship at Cohasset are from Thoreau, *Cape Cod,* 7–9.

Pilot whales . . . are small, toothed whales . . . Pilot whales feed mainly on squid and have been kept in captivity and trained to perform. *Encyclopaedia Britannica CD '97.*

. . . the beach "was strewn, as far as I could see with a glass . . ." Thoreau, *Cape Cod,* 114. The emphasis on the blood and stench of the carcasses continues a theme of *Cape Cod* with which Thoreau began the book. His travels to the Cape began with observations of a shipwreck, including bloated human bodies that had floated ashore and personal belongings of the dead littering the beach. Perhaps, as suggested by Donna Brown, an expert on land use as well as a Thoreau scholar, Thoreau's emphasis in *Cape Cod* on death and destruction on the beach, on ugly scenes, may be part of his ironic humor, given that *Cape Cod* may be seen as an "anti-travel" travel book.

At the time Thoreau was writing the book, steamships and railroads were making tourism accessible to the middle classes, and the genre of travel guides was developing. Most of these early travel guides were superficially enthusiastic about everything and not, from Thoreau's perspective, realistic. With this background, Thoreau's emphasis on blood and guts in *Cape Cod* may be read as simply an attempt to shock the reader. However, in the rest of the book, Thoreau writes with great interest about the people and the dunes and the effect of the sea on him in a manner that is consistent with his other travel writing. My choice, therefore, is to take Thoreau at his word regarding his attitude toward the hunting of the whales.

. . . the decomposing flesh of the whales "might be made into guano . . ." Ibid. Some may take Thoreau's suggestion to use the stinking carcasses as manure as a product of his wry humor, and this might indeed be the case. However, here I take the statement at face value.

"They get commonly a barrel of oil, worth fifteen or twenty dollars, to a fish." This and subsequent quotations from Thoreau about the hunting of the pilot whales in Thoreau, *Cape Cod,* 112–114. Thoreau also noted that some of the excitement was the result of competition among the boaters because "if they succeed in driving them ashore each boat takes one share, and then each man, but if they are compelled to strike them off shore each boat's company take what they strike."

He quoted Timothy Dwight, . . . Timothy Dwight (1751–1817), a descendant of an early Puritan, was one of New England's best-known Congregationalists during the eighteenth century and was president of Yale University at the time of his death. He was a good model for Thoreau in that he was not only a lecturer and author of a travel book but also well known for his farming and his knowledge of music and literature. His *Travels in New England and New York* was a major work of the period, published posthumously in four volumes (1 and 2 in 1821, 3 and 4 in 1822) by S. Converse in New Haven, Connecticut. Information presented here is from the introduction to the reprint edition: Timothy Dwight, *Travels in New England and New York,* edited by Barbara M. Solomon (Cambridge, Mass.: Harvard University Press, Belknap Press, 1969).

"Considering how this State has risen and thriven by its fisheries, . . ." Thoreau, *Cape Cod,* 167.

"There is naked Nature," Ibid., 147–149. However, with forest resources, Thoreau differentiated between use and overuse of trees. That he did not do so with blackfish may be a result of his lack of experience with the ocean and its resources as well as a reflection of the cultural context of his time regarding natural resources. Forests and fields were Thoreau's main focus and occupied most of his time. His three visits to Cape Cod were among his major experiences near and on the ocean and were the primary basis for his writing on this topic.

The rationalist would, as Thoreau did, discuss the use of seaweed and dead pilot whales as fertilizer; . . . Thoreau's attitude toward the whales reflects a strong difference from the attitudes of modern animal rights advocates. Often in society today, the welfare of animals and the well-being of the environment—of ecosystems and the biosphere—are taken to be the same. The saving of an individual whale stranded on a beach is written about in newspapers as an environmental topic. Many people assume that what is best for individual animals, such as beached whales, is best for the entire environment. This, however, is not the case. The difference becomes apparent when one considers the mechanisms of biological evolution and population dynamics. From the viewpoint of a whale, what is "best" evolutionarily is to produce as many offspring as possible. If whales are sentient beings, as some believe, or even are capable of suffering, then what is best for an individual whale is a long life, much food and sex, and little suffering. What is best for the individual leads to rapid population growth. And, as has been well known since the time of René Descartes and Thomas Malthus, a population that grew, unopposed, at the exponential rate of many species would soon overrun the earth, destroying ecosystems and eventually the biosphere.

The following example might help explain the difference between animal welfare and environmental issues. A friend of my daughter collects wild kittens from parks in San Francisco and tries to find homes for them. Her actions are motivated by a sympathy for individual animals—a kind, gentle motivation. Yet a program such as this, if continued for a long time, would lead to an abundance of cats, which could pose a problem to the environment. The goals of animal rights movements may therefore differ from those of environmentalism in general and of the biosphere-oriented deep ecology movement.

British ecologist, Sidney Holt . . . The discussion about Sidney Holt is based on my conversations with him over many years; the interpretation of his reason for opposing all whaling is my own, based on those discussions.

A similar debate was engaged in and a similar conclusion reached by wildlife experts involved in the question of the harvest of elephant ivory. Some, because of their animal rights convictions, wanted to protect elephants. Others were in favor of a reasonable harvest of elephant ivory, arguing that if the elephants provided benefits to the human inhabitants of their habitats, there would be more local concern with the persistence of these animals. However, uncontrolled poaching of elephants grew to such an extent that an international agreement was reached through the Convention on International Trade in Endangered Species (CITES) for a worldwide ban on international trade in ivory. This decision was the result of a rational analysis of the failure of past attempts to control rather than prevent harvest.

"the ocean is a wilderness reaching round the globe, . . ." Thoreau, *Cape Cod,* 148–149.

. . . Yankee whalers were on the high seas, making two- and three-year voyages from nearby New Bedford, . . . The history of hunting of the bowhead whale presented here is from J. R. Bockstoce and D. B. Botkin, "The History of the Reduction of the Bowhead Whale *(Balaena mysticetus)* Population by the Pelagic Whaling Industry, 1848–1914," in *International Whaling Commission Reports,* Special Issue 5 on historical whaling records, edited by M. F. Tillman and G. P. Donovan 1983); and D. A. Woodby and D. B. Botkin, "Stock Sizes Prior to Commercial Whaling," chap. 10 in *The Bowhead Whale,* Special Publication 2, edited by J. J. Burns, J. J. Montague, and C. J. Cowles (Lawrence, Kans.: Society for Marine Mammalogy, 1993), 287–407. See also J. R. Bockstoce, *Whales, Ice, and Men: The History of Whaling in the Western Arctic* (Seattle: University of Washington Press, 1986).

Some scientists in Thoreau's time did see beyond the vastness of the oceans, . . . For example, T. E. Thorpe provided one of the first measurements of atmospheric carbon dioxide concentration, in *Journal of the Chemical Society of London* 5 (1867): 189, and Sir Charles Lyell (1797–1875) discussed the possibility of large-scale effects of life on chemical cycles. Although he rejected the idea, he cited those who believed it. I discuss this in Daniel B. Botkin, *Discordant Harmonies: A New Ecology for the Twenty-First Century* (New York: Oxford University Press, 1990).

. . . Jean Louis Rodolphe Agassiz was one of the discovers of continental glaciation. The discussion here of the discovery of continental-scale glaciation is from chap. 7 of Botkin, *Discordant Harmonies.* The idea first arose in 1815 when a Swiss peasant, J. P. Peeraudin, suggested to a Swiss civil engineer, Ignaz Venetz-Sitten, that some of the features of mountain valleys, including boulders and soil debris, were the result of glaciers that in a previous time had extended down the slopes beyond their present limits. Impressed with these observations, Venetz-Sitten spoke before a natural history society at Lucerne in 1821, suggesting that the glaciers had at some previous time extended considerably beyond their present range.

Agassiz traveled to the Alps to study the evidence firsthand, but with a conclusion drawn in advance—he went with "expectation" that continental glaciation could not have occurred. But once he saw the rocks and debris, he was so impressed with the similarity between the deposits produced by the mountain glaciers and the kinds of topography he had seen elsewhere at lower elevations that he formulated a theory of continental glaciation. Observations had conquered his preconceived ideas.

As discussed in chapter 9, glaciers act as giant bulldozers, pushing everything before them. Over the centuries, with variations in climate, mountain glaciers in the Alps have advanced and receded. During colder periods, glaciers advance and push material ahead of them and to their sides, as when a bulldozer blade spills some of its load alongside its path. At its lower extent, the glacier forms an irregular line of hills made of boulders, rocks, pebbles, sand, silt, clay—every kind and size of material. These formations are called glacial moraines. The debris along the sides, called lateral moraines, are made up of similar materials.

. . . Thomas Jefferson had wondered about the changes that had occurred in distant places, possibly involving the entire Earth. Thomas Jefferson, *Notes on the State of Virginia* (originally published by John Stockdae, London, in 1787), quoted in W. Penn, ed., *Notes on the State of Virginia by Thomas Jefferson* (Chapel Hill: University of North Carolina Press, 1982), 31ff.

"We must be contented to acknowledge, . . ." Ibid., 33.

Plutarch . . . wrote in the first century C.E. . . . Plutarch, "Concerning the Face Which Appears in the Orb of the Moon," quoted in Clarence J. Glacken, *Traces on the Rhodian Shore: Nature and Culture in Western Thought from Ancient Times to the End of the Eighteenth Century,* reprint ed. (Berkeley: University of California Press, 1990), 74.

A British geologist, Adam Sedgwick, argued that vegetation was the primary force opposing erosion . . . Quoted in Sir Charles Lyell, *Principles of Geology; Being an Attempt to Explain the Former Changes of the Earth's Surface by Reference to Causes Now in Operation,* vol. 2 (London: Murray, 1832), 190. Lyell's book is generally accepted as the first modern book on the science of geology.

Sir Charles Lyell, one of the fathers of modern geology, . . . The material about Lyell and others concerning possible global effects of life on the environment is from chap. 9 of Botkin, *Discordant Harmonies.*

Mountains and other areas that had been subjected to such uplifting were observed to be generally denuded of soils . . . Lyell concluded that soils are not conserved during these processes or over long time periods. Moreover, the rate of growth of vegetation is not correlated with the abundance of organic debris; peat, for example, is common in the far north, where vegetation growth is slow, and uncommon in the Tropics, where vegetation growth was believed to be greatest. From these lines of evidence, Lyell concluded that vegetation-derived organic matter in the soil could not be an important cause of creation of landforms or prevention of erosion.

"if the operation of animal and vegetable life . . ." Lyell, *Principles of Geology,* 192.

Lyell then generalized from that case to conclude that vegetation can never play an important role in countering erosion. Lyell also rejected the possibility that organic matter represents a large and increasing fraction of the material on the surface of the earth. Organic soil, which he called "vegetable mould," was "seldom more than a few feet in thickness" and "often did not exceed a few inches"; thus, it represented a small quantity of the earth's material. Furthermore, this organic deposit did not appear to have increased in past geologic periods.

For an excellent modern discussion of factors that determine erosion and situations in which vegetation can and cannot play an important role, see R. C. Sidle, A. J. Pearce, and C. L. O'Loughlin, *Hillslope Stability and Land Use,* Water Resources Monograph Series, no. 11 (Washington, D.C.: American Geophysical Union, 1985). Today, of course, vegetation is understood to play a major role on the land in reducing erosion that has adverse effects on uplands and riparian zones (the land along a stream or river affected by water flow). At this much smaller scale (compared with the ocean), vegetation retards erosion and is important in the development of habitat for salmon and trout in streams. This function of vegetation is at a scale of space and time that intrigued Thoreau, unlike global-scale issues, which did not seem to concern him.

. . . discussions about a global environment and a global nature . . . Geologists and climatologists have pursued these questions and arrived at an understanding of plate tectonics, which is, as far as we know, one of the unique features of the earth and not present on any other body in our solar system. Speculations that human activities were changing the environment and therefore nature at a global scale grew during the twentieth century. The possibility that modern civilization might be changing the earth's climate was suggested in 1938 by G. S. Callendar, based on

measurements of carbon dioxide concentration in the atmosphere taken in the nineteenth and twentieth centuries. The twentieth-century measurements were higher than those made in the nineteenth century. Callendar suggested that the difference could be accounted for by the amount of carbon dioxide added to the atmosphere by the burning of coal, oil, and natural gas since the beginning of the industrial revolution. He also suggested that the increase might lead to a global warming. Callendar was attacked by his scientific colleagues for this suggestion, some of them dismissing the notion simply on the grounds that nineteenth-century scientists could not have done as good a job as scientists in the 1930s and therefore the measurements must have been inaccurate.

Chapter 17. Viewing Our Planet as Nature

"the hours rise up putting off stars . . ." E. E. Cummings, "The Hours Rise Up Putting Off Stars," in *Poems: 1923–1954* (New York: Harcourt Brace, 1954), 42.

"I wish we could sit down and play with these rocks for a while . . ." Quoted in "*Apollo 15* Summary," from the *Apollo Lunar Surface Journal*. Copyright © 1995 by Eric M. Jones. All rights reserved. Available on-line at http://planetscapes.com/solar/eng/apo15.htm.

"some of the grandest thunder which I ever heard . . ." Henry David Thoreau, *The Maine Woods*, designed by Edward Hoagland (New York: Penguin USA, 1988), 324–325.

"The Earth was small, light blue . . ." Unless otherwise indicated, all quotations from astronauts in this chapter are from "Earth from Space," available on-line at http://www.hawastsoc.org/solar/eng/earthsp.htm, which in turn is from Kevin W. Kelley, ed., *The Home Planet* (Reading, Mass.: Addison-Wesley; Moscow: Mir Publishers, 1988).

. . . in that ancient time, some bacteria species evolved that could live in a high-oxygen atmosphere. Free oxygen made possible more efficient and rapid use of energy. A low rate of energy use by the more ancient cells of an oxygenless biosphere required a high surface-to-volume ratio for rapid diffusion of food and chemical elements into cells and for elimination of waste outward. Because surface-to-volume ratio decreases with size, early life-forms had to be small and could not have three-dimensional structures in which cells were surrounded by other cells. They could not be complex organisms with highly differentiated cells and organs. They could not move quickly.

With the presence of free oxygen in the atmosphere, oxidation reactions became possible, enabling organisms to use energy through the process of respiration, an enzymatically controlled "burning" of organic compounds. It then became possible for organisms to evolve complex three-dimensional structures.

. . . our species may have caused some major extinctions . . . Paul S. Martin, *The Last Ten Thousand Years* (Tucson: University of Arizona Press, 1963); P. S. Martin and

C. R. Szuter, "War Zones and Game Sinks in Lewis and Clark's West," *Conservation Biology* 13, no. 1 (1999): 36–45.

They argue that the baseline conditions for land planning should be those of about 10,000 years ago . . . P. Martin and D. Burney, "Bring Back the Elephants," *Wild Earth* 9, no. 1 (spring 1999): 57–64.

The third type of biosphere stage that I believe is in the minds . . . Suppose we were to say that we would like a world that in general maintains members of all the major forms of life, including bacteria, animals, plants, and fungi. Would pilot whales be necessary? The fossil record suggests that there have been long periods during which many kinds of animals and plants have existed without the lesser whales. Most great whales have been brought close to extinction, and we know of few effects on other species when this has occurred. In the modern era, the minimum population of most whale species was reached sometime during the twentieth century. Thus, the answer appears to be no, pilot whales would not be necessary in such a biosphere. One rationale for their conservation is economic (although this economy is in the past, not the present, and is given here only to put the argument in Thoreauvian terms). If we want to sustain the jobs of whalers, we need to sustain the whales, and harvesting would therefore be justified only if it were sustainable. But there is a leniency here not present in the diversity argument: if whales are not especially important to people except as an economic entity, and if this economic benefit is marginal, then the demise and extinction of whales is a minor problem, one that could occur with little or no consequence for human physical needs and perhaps even for human spiritual needs.

Thoreau did not seem to find blackfish of any spiritual value. Civilizations had not arisen from the sea. The scent of the gazelle was not on them. The smell of the sea is not the same as the scents of the deep organic soils of forests, and it is from the latter that civilization arose. Here, a Thoreauvian would have a different argument for sustaining deer than for sustaining blackfish, for the land and its life than for the ocean. But there are other kinds of organisms for which the answer could be yes, including the nitrogen-fixing bacteria mentioned earlier. Some species appear to be more important than others in sustaining the biosphere.

This brings us to the concept of a keystone species, whose primary significance is that its removal would lead to the extinction of many other species or to the cessation of some primary, necessary condition for the persistence of a specific ecosystem. In support of the existence of keystone species, the best case in point is the sea otter, which I discuss in Daniel B. Botkin, *Discordant Harmonies: A New Ecology for the Twenty-First Century* (New York: Oxford University Press, 1990), and elsewhere, and which many others have discussed. It is necessary to return to this example because of the 1.5 million species on the earth, this is the one for which the strongest argument can be made in favor of the keystone species idea.

As described by J. A. Estes and J. F. Palmisano in "Sea Otters: Their Role in Structuring Nearshore Communities," *Science* 185 (1974): 1058–1060, sea otters feed on sea urchins. Sea urchins in turn feed on the bases of kelp, a giant brown alga that creates, in waters just offshore, a kind of ocean "forest" that many fish and shellfish either live their entire lives in or use as a reproductive habitat. Estes and Palmisano's study of islands off Alaska show that where there are sea otters, there are few sea urchins but dense kelp stands and, therefore, habitat for many other creatures. Where there are no sea otters, there are many sea urchins but little kelp and therefore little habitat for species that require it. Sea otters are thus necessary for the persistence of many other species, and therefore they are a keystone species. Pilot whales, in contrast, are pelagic (open-ocean) feeders, compete with many other vertebrates, and if lost would be replaced.

The most impressive counterexample to the keystone species concept is the American chestnut. Informally, it would seem that species that are most abundant in an area or otherwise dominate a particular ecosystem or landscape would be keystone species in the sense described for sea otters. Before the twentieth century, the American chestnut was such a species in the eastern deciduous forests of North America—the forests of chestnut, oak, and hickory that extended from New Jersey down the Piedmont of the Appalachian Mountains to Georgia. With the introduction of a fungal blight from Europe, the American chestnut was rapidly eliminated as a tree (persisting from root sprouts as an occasional single stem, reaching up as far as twenty feet and then dying back). To the best of our knowledge, no other species became extinct or even underwent a major decline as a result of this loss. On wetter sites, red maples became more abundant; on drier sites, oaks became more abundant. Redundancy of function was important to the continued functioning of these forested ecosystems. Species were waiting in the wings to supplant the chestnut.

Here, we run into a central irony of the keystone species concept. Let us do an imaginary experiment in which new fungal blights are introduced one by one into the eastern deciduous forests, each causing the extinction of a major tree species. Suppose that white oak goes first, then black oak, then red oak, then red maple. Eventually, no tree species will be available to grow in certain habitats. If the oaks undergo their demise first, drier sites will lack trees. Finally, only one tree species will be left. With its demise, the forest will disappear. Having eliminated redundancy, we have created a keystone species.

From this, we see that an ecosystem with only one species to perform an important task is vulnerable. Redundancy is valuable for ecosystems, just as the presence of two engines, each capable of keeping a plane aloft, is a safety factor in air travel. But when we have redundancy, we do not have keystone species. The better the situation, the less likely we are to know it to be so. Here, we find a more general argument in favor of conservation of biological diversity, popularized by Paul Ehrlich. It is the qualitative argument that removing species without understanding their function is like randomly removing rivets from an airplane: because there is redundancy of the rivets, removal of the first few will not appear to affect the airplane's structural stability. But eventually, after enough rivets are removed, the plane will fall apart under previously normal stresses. This is the "straw that broke the camel's back" argument. The difficulty with this argument is that we can see its reasonableness when we work upward from one species to a few, but at present we cannot determine the necessary minimum redundancy for the persistence of any ecosystem. The concept of redundancy is not used as a central idea in ecological research. It is not a quantitative factor. It is rarely discussed in this direct fashion, as a metric with value in terms of the persistence of ecosystems. If we were to try to formalize the importance of species diversity for ecosystems, we would need to make redundancy into a formal measure and discuss it in the following terms: By how much does the removal of a redundant species *x* increase the probability of failure of the entire ecosystem? Probability must be stated in terms of the persistence of the ecosystem for some specific time period. To my knowledge, nobody pursues this question in the science of ecology. Once again, we are back with Thoreau on the ice of Walden Pond, but with people speculating about the value of biological diversity rather than the depth of the pond. What we need is a clear formulation of the problem and quantitative information; what we have instead is a simple statement: lacking other knowledge, it is prudent to keep all the species that are now on the earth. The usual statement is: given the rapid rate of extinctions, an increase in this overall rate since the scientific-industrial revolution compared with earlier rates—or, better, compared with rates prior to the arrival of *Homo sapiens* on the earth—it is all the more prudent to hold on to what we have.

From a commonsense point of view, this is a defensible position. But over the long run, pushed up against the needs of a human population of 6 billion and growing, pushed up against the desires of people with money and power, against nations with vastly different political, religious, and cultural drivers, this argument will need to have much more teeth in it. It must be delved into much deeper, must be the rock at the end of Thoreau's rope on the ice at Walden Pond.

The earliest known alphabet . . . See John Noble Wilford, "Finds in Egypt Date Alphabet in Earlier Era," *New York Times*, 14 November 1999.

. . . it is therefore dangerous to equate maximum species diversity with the maximum likelihood of persistence of the biosphere. As an analogy, consider the design of an airplane, in which the airplane itself is the system of interest. Suppose we were to argue that the most diverse design would be the safest: the more engines, the better; the more redundant the control systems, the better; the more redundant in function but different in operation the instruments, the better. But this argument rapidly leads to an absurd design. The plane becomes dangerously heavy, and its complexity leads to control problems. This was one of the driving factors in the transition from piston engine aircraft to jet aircraft. At the time of development of the jet engine, some engineers believed it would never be sufficiently fuel efficient to be useful and instead urged an increase in the number of pistons per piston engine and the number of engines. But the design of these ever more complex engines became impractical. Each additional engine represented an additional source of fire or other catastrophic failure that could threaten the entire system. A more successful design would have some degree of redundancy and diversity, but not the maximum. There is an optimal level of complexity and therefore of diversity for this kind of machine.

With this analogy in mind, consider the much more complex biosphere and the hypothesis that maximizing biological diversity will maximize its persistence. That the longest-persisting biosphere was the earliest, bacterial, with very low diversity, rejects this hypothesis. The theoretical comparison with an airplane design suggests that it is also unlikely. Therefore, it is dangerous ground to tread to try to argue that there is a scientific basis for maximization of biological diversity, or that one can expect science to prove this because it seems so obvious or so desirable on other grounds.

The problem here is "linear" thinking of a simple kind. If two are good and three are better, then the more, the better. There is a discussion that is not possible in this context but is relevant. It has to do with the ways in which life affects the rest of the biosphere. On land, vegetation and bacteria have some of the most significant effects on the rest of the earth's life-supporting system. We know today that climate is influenced by four characteristics of vegetation cover: color and reflectance; water evaporation; release and uptake of chemical elements and compounds, including greenhouse gases; and surface roughness. Surface roughness is just what it would seem: winds are slowed by a rough surface, and tall trees are a rougher surface than bare soil. Thus, vegetation cover can moderate wind speed and therefore influence weather. In the oceans, plankton, made up of minute animals and blue-green algae (photosynthetic, nitrogen-fixing bacteria), affects the exchange of oxygen and carbon dioxide, and the nitrogen-fixing bacteria affect the nitrogen cycle. Large areas with life on them affect the earth's atmosphere, but the effect, from a global perspective, could be mimicked by a designed and heavily modified vegetation cover on the land. Similar information exists for other forms of life, but a detailed examination of this is beyond the scope of the present discussion. Readers interested in the topic are referred to B. A. Skinner, S. Porter, and D. B. Botkin, *The Blue Planet* (New York: Wiley, 1999).

. . . *Thoreau's approach to knowledge* . . . The essential elements of what I have referred to as Thoreau's approach to knowledge are as follows: willingness to examine both the tangible (physical) and intangible (creative, inspiring, spiritual, aesthetic, religious) aspects of human contact with nature; examination of how the tangible and the intangible affect each other and under what conditions one or the other should be relied on; application of the scientific method, including use of quantitative information wherever possible; detailed examination of natural phenomena but as much as possible without "expectation"—that is, without a prior conclusion drawn or an ideology imposed on the subject; willingness to consider the ideas of others as starting points—hypotheses—whether these ideas come from professional experts, like Emerson and Agassiz, or experiential experts, like the Wellfleet Oysterman and the Old Woman on Cape Cod; and concern with the relationship between civilization and nature as well as that between the individual and nature, so that questions are examined within this general context.

Regarding a moral point of view about the biosphere, there is a distinction between the statement "Please, let nature alone," and the statement "You must leave nature alone." The former can be taken as a statement made with humility, a request that we not tread where we do not know, a request that we recognize that an understanding of the complex system that is the biosphere yet eludes us. The latter is a normative statement. It is a strong policy statement that presupposes possession of special knowledge, either spiritual or physical, by the speaker. Where it is presumed to be based on knowledge of how the biosphere works, it is an assertion that we are in possession of knowledge that in fact eludes us. It is therefore empty of content and to be avoided.

Chapter 18. Conserving Mono Lake: Walden Pond as Metaphor

"Now the elements of the art of war are first . . ." Sun-tzu, *The Art of War,* originally written circa 400 B.C.E. (New York: Oxford University Press, 1971), 20.

"The islands in [Mono Lake] being merely huge masses of lava . . ." Mark Twain, *Roughing It* (New York: Penguin Classics, 1985), 276–277.

The National Audubon Society helped raise funds . . . John Hart, *The Mono Lake Battle and the California Water Future* (Berkeley: University of California Press, 1996).

Originally, the California Department of Fish and Game was to conduct the study . . . The information about the scientific study of Mono Lake are from my personal experiences and from: Botkin, D. B., W. S. Broecker, L. G. Everett, J. Shapiro, and J. A. Wiens, 1988, The Future of Mono Lake, California Water Resources Center, University of California, Riverside, Report #68; Botkin, D. B. and J. A. Wiens, 1988, Mono Lake: Solving An Environmental Dilemma, The World and I, 3(5): 198–205; and Wiens, J. A., D. T. Patten, D. B. Botkin, 1993, Assessing Ecological Impact Assessment:: Lessons from Mono Lake, California, Ecological Applications 3(4): 595–609.

Chapter 19. Cities, Civilization, and Nature

"Earth has not anything to show more fair . . ." William Wordsworth, *Upon Westminster Bridge,* A. Quiller-Couch, ed., *The Oxford Book of English Verse, 1250–1900* (Oxford: Clarendon Press, 1912), 584.

"We require an infusion of hemlock spruce or arbor-vitae in our tea." Henry David Thoreau, "Walking," in *The Natural History Essays,* edited by Robert Sattelmeyer (Salt Lake City: Peregrine Smith Books, 1980), 113.

"I think that I love society as much as most . . ." Henry David Thoreau, *Walden,* edited by J. Lyndon Shanley (Princeton, N.J.: Princeton University Press, 1971), 140.

"I had more visitors while I lived in the wood than at any other period of my life; I mean that I had some, . . ." Ibid., 143–144.

We must focus on cities as nature. Much of the material that follows appeared originally in D. B. Botkin and C. E. Beveridge, "Cities as Environments," *Urban Ecosystems* 1, no. 1 (1997): 3–20.

In 1950, there were just two cities that, along with their suburbs, had populations of 10 million. World Bank, *World Development Report* (Oxford: Oxford University Press, 1984).

Roderick Nash, a well-known environmental historian, wrote recently that he hopes that in the future there will be an "urban implosion" . . . R. F. Nash, "Island Civilization: A Vision for Planet Earth in the Year 2992," *Wild Earth* (winter 1991–1992): 2–4.

"Every tree sends its fibres forth in search of the Wild. . . ." Thoreau, "Walking," 112.

"The African hunter Cumming . . ." Thoreau, "Walking," 112–114.

"Our village life would stagnate . . ." Thoreau, *Walden*, 317.

"Even the oldest villages . . ." Henry David Thoreau, *A Week on the Concord and Merrimack Rivers*, edited by Carl F. Hovde, William L. Howarth, and Elizabeth H. Witherell (Princeton, N.J.: Princeton University Press, 1980), 171.

. . . the Widneyville developers . . . B. A. Botkin, *A Treasury of Western Folklore* (New York: Crown, 1978).

This kind of climate is rare, occurring on less than 2 percent of the Earth's surface . . . L. W. Lenz and J. Dourley, *California Native Trees and Shrubs: For Garden and Environmental Use in Southern California and Adjacent Areas* (Claremont, Calif.: Rancho Santa Ana Botanic Garden, 1981), 232.

In Two Years before the Mast, *. . .* Richard Henry Dana, *Two Years before the Mast: A Personal Narrative of Life at Sea* (New York: Harper, 1840; reprint, New York: Penguin Books, Signet Classics, 1964), 182.

the town as lying "on a low plain . . ." Ibid., 57.

"The town is finely situated . . ." Ibid., 58. Regarding the fire in Santa Barbara, Dana continued: "The fire was described to me by an inhabitant as having been a very terrible and magnificent sight. The air of the whole valley was so heated that the people were obliged to leave the town and take up their quarters for several days upon the beach."

"These trees are seldom more than five or six feet high . . ." Ibid., 146.

"upper parts of the coast, and in the interior." Ibid., 149.

How does one bring nature that people actually enjoy to the cities of such a region? For more information about the biological diversity of the region and the application of Olmsted's ideas to vegetation plantings in southern California, see Botkin and Beveridge, "Cities as Environments."

Olmsted was born in Hartford, Connecticut, on April 26, 1822, . . . C. E. Beveridge and Paul Rocheleau, *Frederick Law Olmsted: Designing the American Landscape* (New York: Rizzoli, 1995), 10. Beveridge notes that Olmsted was the eighth-generation descendant of James Olmsted, one of the founders of Hartford in 1636.

This set a new direction for Olmsted's career as America's first and greatest landscape architect. Beveridge and Rocheleau, *Frederick Law Olmsted.*

Olmsted wrote that vegetation in cities plays social, medical, and psychological roles. F. L. Olmsted to George W. Elliott, 28 April 1890, letterpress book A7, 373–374.

In the science of ecology, scientists and practitioners have generally shown little interest in urban environments . . . Among the changes taking place in the science of ecology to increase the emphasis on urban environments are two new projects of the Long-Term Ecological Research program (supported by the National Science Foundation) focused on cities, one for Baltimore and one for Phoenix; a new journal, *Urban Ecosystems;* and many ecological restoration activities taking place in or near cities. Examples of the latter are the restoration of prairies near Chicago and the reconnection of Portland, Oregon, to the Willamette River, accomplished by moving a major interstate highway, I-5, across the river.

In the history of city planning . . . The material that follows regarding the history of city planning is from Daniel B. Botkin and Edward A. Keller, *Environmental Science: Earth as a Living Planet,* 2nd ed. (New York: Wiley, 1997), and Botkin and Beveridge, "Cities as Environments."

Olmsted used the skills in both engineering and landscape design . . . Since the time of Olmsted's work, a number of researchers have found that vegetation in urban areas can indeed provide emotional and psychological benefits. For example, one study found that visitors to Detroit's Belle Isle park, most of whom were from low-income, inner-city areas, experienced significant reduction in stress. J. Huang, R. Ritschard, N. Sampson, and H. Taha, "The Benefits of Urban Trees," in H. Akbari, S. Davis, S. Dorsano, J. Huang, and S. Winnett, *Cooling Our Communities: A Guidebook on Tree Planting and Light-Colored Surfacing,* ISBN 0-16-036034-X (Washington, D.C.: Environmental Protection Agency, Office of Policy Analysis, 1992), 27–42.

"I have given the matter of pleasure grounds for San Francisco some consideration . . ." F. L. Olmsted to William Hammond Hall, 5 October 1871, Frederick Law Olmsted Papers, Manuscript Division, Library of Congress, Washington, D.C.

"Cutting yourself completely clear of the traditions of Europe and the East . . ." Frederick Law Olmsted to William Hammond Hall, 1874, in "Extracts from Letters Found of Record in the Files of the Board of Park Commissioners," in *The Development of Golden Gate Park and Particularly the Management and Thinning of Its Forest Trees* (San Francisco, 1886), 31–32.

The idea of irrigated lawn as, properly, a communal possession was repeated in two of Olmsted's proposals. Frederick Law Olmsted, "Report upon a Projected Improvement of the Estate of the College of California, at Berkeley, Near Oakland," "The Papers of Frederick Law Olmsted" 5: 566.

. . . Olmsted proposed four principles of semi-arid landscape design . . . Olmsted's son carried on the task of devising an approach to the use of vegetation in the region, a process his father had initiated fifty years before. Their combined work suggests how vegetation can be used to improve cities in the American West. Their plans suggest that vegetation can be planted to meet human social and psychological needs and that irrigation can be used carefully and wisely to enhance the city environment.

What we need to do is integrate this philosophy and these basic design attributes with modern environmental needs and the new understanding of ecological systems. In southern California in recent years, the major approach to the combined problems of a limited water supply and the need for a pretty landscape has been the Xeriscape (from *xeric,* derived from a Greek word meaning "dry") method of gardening with plants adapted to dry country—deserts and semideserts. Botanists and plant ecologists call such plants xerophytes (*xero-* meaning dry, *-phyte* denoting a plant with a specified characteristic). This would seem the right approach for Los Angeles for two reasons. First, it minimizes water use. Second, because southern California, except at high elevations, lies within a semi-arid climatic zone, one might think that only xerophytic vegetation is native. If this is so, Xeriscape landscaping might also serve a biological conservation function.

During the 1980s, Xeriscape gardening became popular. My family used it in our backyard in Santa Barbara, removing a lawn and replacing it with a Mexican tile patio and a circular path of river-rounded pebbles, with xeric flowering plants in the center of the circle, along the arch-shaped edges, and in the background. Some of the plants were native. We planted a native California oak that we obtained from a friend who ran a large tree nursery, and several other oaks "volunteered"—sprouted without our help. We bought a horticultural variety of the native California lilac whose flowers are a brilliant blue rather than the pale whitish-blue of the variety common in the surrounding hills. But many of the plants had been introduced into California from around the world. These were selected because they require less water and have brighter blossoms or more graceful forms than native vegetation. We planted an Australian pepper tree to shade the patio because of its graceful, willow-like form, cascading branches, and fine leaves that provide shade. This backyard, on more or less level terrain, has worked for us. It is more decorative than a lawn, and we walk on the paths more than we had previously walked on the grass.

The firefighters stationed at one of Santa Barbara's firehouses built a demonstration Xeriscape garden for people to visit and see what they might do similarly at home. Friends of ours lived in the house nearest to this fire station, just down a slope from the garden. Everything was fine until a long drought ended and there was a heavy rain. The Xeriscape garden gave way, and a good portion of its soil and vegetation ended up in our friend's living room. Thus, in a countryside of unstable soils, Xeriscape gardens clearly are not the ultimate answer. Xeric plants are shallow rooted or widely spaced and provide little structure to hold back erosion. My friend's story illustrates the limitations of a single-factor approach to nature and to environmental design.

Part of the parkway's . . . Olmsted's parkways included separate ways for each kind of transportation—pedestrians; people on horses, which have, of course, become people in automobiles; and bicyclists—and sometimes made provision for street railways as well. In addition to creating a green, shaded route through the city, parkways provided neighborhood green space. As Olmsted described the result: "Thus, at no great distance from any point of the town, a pleasure ground will have been provided for, suitable for a short stroll, for a playground for children and an airing ground for invalids, and a route of access to the large common park of the city, of such a character that most of the steps on the way to it would be taken in the midst of a scene of sylvan beauty, and with the sounds and sights of the ordinary town business, if not wholly shut out, removed to some distance and placed in obscurity. The way itself would thus be more park-like than town-like." F. L. Olmsted to William Dorsheimer, 1 October 1868, in *First Annual Report of the Buffalo Park Commissioners, January, 1871* (Buffalo, N.Y.: Warren, Johnson), 26.

"I see, in the thin snow along by the button-bushes . . ." Henry David Thoreau, *The Writings of Henry David Thoreau,* edited by Bradford Torrey, vol. 11, *Journal* (Boston: Houghton Mifflin, 1906), 376 (25 December 1858).

"You cannot go out so early but you will find the track of some wild creature. . . ." Henry David Thoreau, *Journal,* edited by Bradford Torrey (Boston: Houghton Mifflin; Cambridge, Mass.: Riverside Press, 1906), 432 (22 December 1852).

"For many years I was self-appointed inspector of snow storms and rain storms . . ." Thoreau, *Walden,* 18.

Chapter 20. Civilization and Nature

"Two men in a skiff . . ." Henry David Thoreau, *A Week on the Concord and Merrimack Rivers,* edited by Carl F. Hovde, William L. Howarth, and Elizabeth H. Witherell (Princeton, N.J.: Princeton University Press, 1980), 48–49.

Some theologians . . . This was the position taken by some of the Rabbinical scholars at a conference on Judaism and the environment held by the World Center for Religion at Harvard University in 1998.

. . . in coining the term technological imperative. This is a personal communication from Garrett. I believe he has written it somewhere, but it was most interestingly expressed during a conversation we had several years ago.

Before we get around to saving ourselves by controlling the size of our population, . . . Our inability to change our way of thinking is illustrated by the methods in use at present to forecast the growth of the human population. If you think that the understanding of the growth of the human population has passed beyond the old ideas of the balance of nature and of a static biosphere, consider how the calculations of the growth of our population are made. Projections in use widely, produced by the World Bank and the United Nations, are based on the assumption that the human population will grow along a logistic growth curve to some steady-state size that will persist indefinitely. All the major projections used in major political debates about the growth of our species use this method. Assumed to grow according to this curve, the human population is implicitly assumed to be part of a balance of nature, growing according to the assumptions of that old idea. The only speculation allowed is when the inflection point in the logistic curve will occur. That determines the ultimate limit of human abundance. Thus the United Nations provides three forecasts that vary in the expected value of that final number. Our approach to forecasts of our own population lag behind ideas as much as the repair of bridges lagged behind their decay.

. . . I was struck by a statement made by Roger Sedjo, an economist who specializes in forestry issues. The analysis of the use of plantations to meet world timber needs was published in R. A. Sedjo and D. B. Botkin, "Using Forest Plantations to Spare the Natural Forest," *Environment* 39, no. 10 (1997): 14–20.

If we were to try to produce all the timber traded from this kind of forest, we would need 20 to 50 percent of the world's forestland. This would be about 750 million to 1.5 billion hectares (1.9–4.0 billion acres), or about 2.9–6.3 million square miles, of forestland.

. . . he found nature "something savage and awful, though beautiful" Henry David Thoreau, *The Maine Woods,* 2nd ed., edited by Joseph J. Moldenhauer (Princeton, N.J.: Princeton University Press, 1973), 70.

Acknowledgments

This book began with a casual conversation with documentary filmmaker Ted Timreck. We were trying to get funding for a film about Lewis and Clark, based on my previous book, *Our Natural History: The Lessons of Lewis and Clark*. Ted said, "Why don't you do a book like that about Thoreau—then you'd have done one for the east and one for the west." At the time, it seemed an obvious and straightforward thing to do. I had read Thoreau's works in the early 1970s and thoroughly enjoyed *The Maine Woods* and *Cape Cod*. But when I became involved in the work, I rediscovered that Thoreau was much more complex a subject than were the travels of Lewis and Clark. While Lewis and Clark were great heroes and leaders, they excelled at describing the countryside. Thoreau, instead, told us what it was like to be there. The boreal forest of Maine in which he traveled had few tree species, and Thoreau did not bother to describe the exact location and the botanical situation of every site; in fact, sometimes it is hard to tell just what day he is describing. But I began to realize that Thoreau had messages for us, which I have tried to portray in this book. Ted and I traveled some of Thoreau's paths together and developed a draft script about Thoreau whose echoes sound through this book. So I thank Ted very much for sending me off on this journey, even if it was different and much harder than I had thought.

Every part of this book in every draft was read carefully by my sister, Dorothy B. Rosenthal, and I am much indebted to her for her direct and valuable comments, as well as many other suggestions about things to read and our many discussions during our lives. As I mentioned in the dedication, she has been my best and harshest critic and this book would be much poorer without her contributions.

Dan Sayre, editor in chief at Island Press, spent many hours reading, reviewing, critiquing, and discussing the manuscript with me. Without our vigorous and stimulating discussions, this book would be less clear and cogent. He remained throughout a strong supporter as well as an excellent critic of the work. Chuck Savitt, president of Island Press, supported the idea of this book from its inception and also contributed many useful discussions about conservation in America. I also thank Bill LaDue, production director at Island Press, and his staff for careful editing and preparation of the manuscript. I thank David Bullen for his design and Joyce Powzyk for her illustrations.

Ever since I met Charles Beveridge in graduate school at the University of Wisconsin, he and I have discussed civilization and nature, and he has been a wealth of sources of references, ideas, and insights about the history of people and nature. Charlie's insights have greatly influenced the chapter about cities, and he has kindly allowed me to use material from a paper he and I wrote together for the city of Los Angeles. As editor of the Olmsted papers, Charlie's knowledge of Olmsted greatly influenced and set the direction for my discussion of the father of landscape architecture.

Lee Talbot, one of our time's most important and effective conservationists, has been for many years a friend and colleague with whom I have also learned a great deal about people, nature, and the politics of conservation. He, along with Charlie Beveridge, read drafts of this manuscript and made important suggestions to which I have tried to respond.

Parker Huber, who has walked every walk of Thoreau's and canoed every waterway that Thoreau did, guided me and Ted Timreck through a part of Thoreau's Maine woods and Thoreau's Cape Cod. But more than that, with his marvelous empathy for Thoreau, he helped guide me in my thoughts about this subject, at the same time becoming a good friend.

Elizabeth Witherell, editor in chief of *The Writings of Henry D. Thoreau* at Northern Illinois University, has been a steady source of specific information about Thoreau, but much more important, for the past four years, Beth and I have talked often about Thoreau and the meaning of his work. Her concurrence with my interpretations has been of extraordinary importance to me in allowing me to go forward with this project.

Bradley P. Dean, director of the Media Center at the Thoreau Institute, provided information about specific events in Thoreau's life and works. His general perceptions of Thoreau and enthusiasm for Thoreau and works about him played an important role as this book developed.

Professor Mac Hunter of the University of Maine put me in direct contact with Carl Haag, the Maine forester so important in part of this book, took me on the hike to Reed Pond, arranged for me to speak at the University of Maine and test out my ideas, and carefully and thoughtfully reviewed an early draft of this manuscript. I hope the final product lives up to the quality of his help.

Professor Matthew Sobel of Case Western Reserve University has cooperated with me for many decades, since we met while on the faculty of Yale University. His clear thinking, knowledge of mathematics, and great judgment about many subjects have helped steer me along a straighter path than I would have otherwise followed.

Carl Haag, now of Plum Creek Corporation and a longtime Maine forester, helped me greatly with the experiences of himself and John M. Hagan III, and with an understanding of forestry issues in Maine. His thoughtfulness is of great value. Buzz Caverly, director of Baxter State Park, kindly reviewed the chapter on that park and provided insight into the effects of Baxter's deeds of gift on the work of the director.

Karen Redden worked for me for five years as a research assistant; she tracked down much material for this book, more often obscure than not, and did so with great enthusiasm. She helped keep material organized so that it looked slightly better than what James Thurber referred to as the material that he let "fall behind his roll top desk."

Playwright and professional writer and editor Joan Melcher read several drafts of this manuscript and made important suggestions. She has been a long-term reader and critic of my work, always to my benefit.

Susan Day, who worked as the administrator of The Center for the Study of the Environment, read and helped with much background material that is imbedded in the discussions in this book.

Many of my professional colleagues discussed ideas in this book with me over the years and participated in projects I discuss, including those about Mono Lake and salmon. These include Wallace Broecker, Columbia University; Ken Cummins, Humboldt State University; Thomas Dunne, University of California, Santa Barbara; Lorne Everett, University of California, Santa Barbara; Roger Sedjo, Resources for the Future; Joseph Shapiro, University of Minnesota; and Thomas Wiens, Colorado State University.

James Brown, Oregon state forester, and Richard Pfilf, USDA Forest Service (retired), provided valuable insights into the conservation and management of forest resources. Insights into the importance of the fossil record have been helped by work with Lynn Margulis, University of Massachusetts; Brian Skinner, Yale University; Stephen Porter, University of Washington; and Bruce Tiffney, University of California, Santa Barbara.

Carol Mann, my literary agent, helped by supporting the idea that I write about topics of importance to me, and she smoothed the way during the entire process of the development of this book.

Before she succumbed to cancer, my wife, Erene P. Botkin, enthusiastically supported this project and remains an influence on me and my work.

I also acknowledge the work of Lloyd Simpson and Robert Nisbet in the analysis of information related to forestry, salmon, and other environmental issues and ecological research.

Finally, and most important, I thank Jane O'Brien for her companionship, for her support for my work, and for living through the process of the writing of this book.

Permission Statements

Index